Current Developments in Mathematics 1996

Edited by

Raoul Bott
Arthur Jaffe
David Jerison
George Lusztig
Isadore Singer
S.-T. Yau

International Press
www.intlpress.com

Current Developments in Mathematics, 1996

ISBN 978-1-57146-147-6

Paperback reissue 2010. Previously published in 1997 under ISBN 1-57146-035-7 (clothbound).

Typeset using the LaTeX system.

Preface

These are the proceedings of the joint seminar by M.I.T. and Harvard on the current developments in mathematics for the year 1996. Established in 1995, this seminar will be continued each year. In 1997 and following years, we will also add a section on open problems.

The organizing committee for the seminar is made up of representatives from the mathematics departments of the two institutions: Raoul Bott, Arthur Jaffe, and S.T. Yau from Harvard; and David Jerison, George Lusztig, and Isadore Singer from M.I.T.

We would like to thank each of the contributors and to recognize the several institutions without whose participation the seminar would not have been possible: the departments of mathematics at Harvard and MIT for their financial aid and the American Academy of Arts and Sciences for opening their facilities to us.

Editors

Contents

Automorphic forms and Lie algebras

Richard E. Borcherds [1]
D.P.M.M.S.,
16 Mill Lane,
Cambridge,
CB2 1SB, England
E-mail: reb@pmms.cam.ac.uk
Home page:
http://www.dpmms.cam.ac.uk/ reb

Introduction

This paper is mainly an advertisement for one particular Lie algebra called the fake monster Lie algebra. The justification for looking at just one rather obscure object in what is supposed to be a general survey is that the fake monster Lie algebra has already led directly to the definition of vertex algebras, the definition of generalized Kac-Moody algebras, the proof of the moonshine conjectures, a new family of automorphic forms, and it is likely that there is still more to come from it.

The reader may wonder why we do not start by looking at a simpler example than the fake monster Lie algebra. The reason is that the fake

[1]Supported by a Royal Society professorship and by NSF grant DMS-9401186.

monster is the simplest known example in the theory of non-affine Lie algebras; the other examples are even worse.

Here is a quick summary of the rest of this paper. The first sign of the existence of the fake monster came from Conway's discovery that the Dynkin diagram of the lattice $II_{25,1}$ is essentially the Leech lattice. We explain this in section 1. For any Dynkin diagram we can construct a Kac-Moody algebra, and a first approximation to the fake monster Lie algebra is the Kac-Moody algebra with Dynkin diagram the Leech lattice. It turns out that to get a really good Lie algebra we have to add a little bit more; more precisely, we have to add some "imaginary simple roots", to get a generalized Kac-Moody algebra. Next we can look at the "denominator function" of this fake monster Lie algebra. For affine Lie algebras the denominator function is a Jacobi form which can be written as an infinite product [K, chapter 13]. For the fake monster Lie algebra the denominator function turns out to be an automorphic form for an orthogonal group, which can be written as an infinite product. There is an infinite family of such automorphic forms, all of which have explicitly known zeros. (In fact it seems possible that automorphic forms constructed in a similar way account for all automorphic forms whose zeros have a "simple" description.) Finally we briefly mention some connections with other areas of mathematics, such as reflection groups and moduli spaces of algebraic surfaces.

I thank S.-T. Yau for inviting me to this conference, U. Gritsch, I. Grojnowski, G. Heckman, and M. Kleber for many helpful remarks about earlier drafts, and the NSF for financial support.

1 The Leech lattice and $II_{25,1}$

In this section we explain Conway's calculation of the reflection group of the even Lorentzian lattice $II_{25,1}$, so we first recall the definition of this lattice. A lattice is called even if the norms (v, v) of all vectors v are even and is called odd otherwise. The lattice is called unimodular if every element of the dual of L is given by the inner product with some element of L. The positive or negative definite lattices seem impossible to classify in dimensions above about 30 as there are too many of them, but the indefinite unimodular lattices have a very simple description: there is exactly one odd unimodular lattice $I_{r,s}$ of any given dimension $r + s$ and signature $r - s$ for positive integers r and s, and there is exactly one even unimodular lattice $II_{r,s}$ if the signature $r - s$ is divisible by 8 and no even unimodular ones otherwise. (Note that the II in $II_{r,s}$ is two I's and not a capital Π!) In particular there is a unique even unimodular 26 dimensional Lorentzian lattice $II_{25,1}$.

The odd lattice $I_{r,s}$ can be constructed easily as the set of vectors

$$(n_1, \dots, n_{r+s}) \in \mathbb{R}^{r,s} \text{ with all } n_i \in \mathbb{Z}.$$

(The norm of this vector in $\mathbb{R}^{r,s}$ is $n_1^2 + \cdots n_r^2 - n_{r+1}^2 - \cdots n_{r+s}^2$.) The even lattice can be constructed in the same way, except that the conditions on the coordinates are that their sum is even, and they are either all integers or all integers $+1/2$. Notice that the vector $(1/2, \dots, 1/2)$ has even norm $(r-s)/4$ because the signature $r-s$ is divisible by 8. For $r=8, s=0$ this is of course just the usual construction of the E_8 lattice. There is a second way to construct $II_{25,1}$: if Λ is the Leech lattice (the unique 24 dimensional even unimodular positive definite lattice with no roots) then $\Lambda \oplus II_{1,1}$ is an even 26 dimensional Lorentzian lattice and is therefore isomorphic to $II_{25,1}$. For the lattice $II_{1,1}$ it is convenient to use a different coordinate system: we represent its vectors as pairs $(m,n) \in \mathbb{Z}^2$ with the norm of (m,n) defined to be $-2mn$, so that $(1,0)$ and $(0,1)$ are norm 0 vectors. We use this to write vectors of $II_{25,1}$ in the form (λ, m, n), where $\lambda \in \Lambda$ and $m, n \in \mathbb{Z}$, where this vector has norm $\lambda^2 - 2mn$.

Conversely Conway showed [CS] that we can run this backwards and give a very short construction of the Leech lattice. If we let

$$\rho = (0, 1, 2, \dots, 22, 23, 24, 70) \in II_{25,1}$$

then $\rho^2 = 0$ and $\rho^\perp/\rho$ is isomorphic to the Leech lattice.

In the rest of this section we describe Vinberg's algorithm for finding simple roots of reflection groups and Conway's application of it. If L is any Lorentzian lattice the norm 0 vectors form two cones and the negative norm vectors are the "insides" of these cones. We select one cone and call it the positive cone C. The set of norm -1 vectors in C forms a copy of hyperbolic space H. (The metric on H is just the pseudo Riemannian metric on $L \otimes \mathbb{R}$ restricted to H, where it becomes Riemannian.) The group $Aut(L \otimes \mathbb{R})^+$ of all rotations of $L \otimes \mathbb{R}$ mapping C to itself acts on the hyperbolic space H, and is in fact the group of all isometries of H. In particular the group $Aut(L)^+$ of all automorphisms of L fixing the positive cone can be thought of as a discrete group of isometries of H. We let W be the subgroup of $Aut(L)^+$ generated by reflections. These reflections can be described as follows: if r is a positive norm vector of L such that $(r,r)|2(r,s)$ for all $s \in L$ then the reflection in $r^\perp$ given by $s \mapsto s - 2r(r,s)/(r,r)$ acts on $L \otimes \mathbb{R}$ and by restriction on H, where it is reflection in the hyper-space $r^\perp \cap H$.

The reflection hyperspaces divide hyperbolic space into cells called the Weyl chambers of W.

Just as in the case of finite Weyl groups the Weyl group W acts transitively on the Weyl chambers. We select one Weyl chamber D and call it the fundamental Weyl chamber. Then the full group $Aut(L)^+$ is the semidirect product of W and the subgroup of $Aut(L)^+$ fixing the fundamental Weyl chamber D. In particular if we can describe D this more or less determines the group $Aut(L)^+$.

Vinberg invented the following algorithm for finding the shape of D. First choose any point c in the fundamental Weyl chamber D. Then find the faces of D in order of their distance from c. Vinberg showed that a reflection hyperplane is a face of D if and only if it makes an angle of at most $\pi/2$ with each face of D nearer c (where the angle between faces means the angle as seen from inside D). This means we can find all the walls of D recursively in order of their distances from c.

It is convenient to rephrase this algorithm in terms of the lattice L rather then the hyperbolic space H. Instead of a point c we choose a vector ρ inside the closed positive cone C. For simplicity we assume that all the roots of W have norm 2 (which is the only case we will use later). Then we replace the distance from a hyperplane to the point c by the height $-(\rho, r)$ of r where r is any root having inner product at most 0 with ρ. Then as before we can find all the simple roots (i.e., positive roots r orthogonal to a face of D) in order of their heights, by observing that a positive root r is simple if and only if it has inner product at most 0 with all simple roots of smaller height.

We can now describe Conway's calculation [C] of the simple roots of the reflection group W of $II_{25,1}$ using Vinberg's algorithm. We choose the vector ρ to be $(0,0,1) \in \Lambda \oplus II_{1,1}$. There are no roots of height 0 because the Leech lattice Λ has no roots. The vectors $(\lambda, 1, \lambda^2/2 - 1)$ for $\lambda \in \Lambda$ are the norm 2 simple roots of height 1 and it is easy to check that they form a set isometric to the Leech lattice Λ.

Conway's amazing discovery was that these are the only simple roots of W. To see this suppose that (v, m, n) with $v^2 - 2mn = 2$ is any other simple root of height $m > 1$. Then (v, m, n) has inner product at most 0 with all the simple roots $(\lambda, 1, \lambda^2/2 - 1)$ for all $\lambda \in \Lambda$ and an easy calculation shows that this implies that $|v/m - \lambda| > \sqrt{2}$ for all λ. But Conway, Parker, and Sloane [CS, chapter 23] showed that the Leech lattice has covering radius exactly $\sqrt{2}$, in other words $\Lambda \otimes \mathbb{R}$ is just covered by closed balls of radius $\sqrt{2}$ around each lattice point of Λ. In particular there is some vector λ of Λ with $|v/m - \lambda| \leq \sqrt{2}$, and this contradiction proves that (v, m, n) cannot be a simple root.

To summarize, this gives a complete description of the group $Aut(II_{25,1})^+$ as a semidirect product $W.G$ where W is the reflection group with Dynkin diagram given by the Leech lattice as above, and where G is the group of automorphisms of this Dynkin diagram. It is not hard to see that G is just the group of all isometries of the "affine" Leech lattice (which is the Leech lattice with the origin "forgotten"), and so is a semidirect product $\Lambda.Aut(\Lambda)$ where $Aut(\Lambda)$ is the double cover $(\mathbb{Z}/2\mathbb{Z}).Co_1$ of Conway's largest sporadic simple group Co_1. So $Aut(II_{25,1})$ can be written as

$$W.\Lambda.(\mathbb{Z}/2\mathbb{Z}).Co_1.$$

This description of Co_1 as the group sitting at the top of $Aut(II_{25,1})$ seems to be the simplest and most natural description of any of the sporadic simple groups.

Given any Dynkin diagram there is an associated Kac-Moody algebra. As the Dynkin diagram of the reflection group of $II_{25,1}$ is so nice, this suggests that the associated Kac-Moody algebra M_{KM} should also be very nice. This turns out to be almost but not quite true: we first have to modify the Dynkin diagram by adding some imaginary simple roots in order to get a nice Lie algebra (which will be the fake monster Lie algebra M, whose maximal Kac-Moody subalgebra is M_{KM}.). Before discussing this we recall some facts about Kac-Moody algebras.

2 Kac-Moody algebras

Suppose that G is a finite dimensional simple complex Lie algebra with Cartan subalgebra H. Then G has a symmetric invariant bilinear form $(,)$ which induces a form on H, which we use to identify H with its dual H^*. The roots $\alpha \in H^*$ are the nonzero eigenvalues of the adjoint action of H on G and are the roots of a finite reflection group W acting on H called the Weyl group of G. The simple roots α of some fundamental Weyl chamber D of W can be identified with the points of the Dynkin diagram of G.

We define the numbers a_{ij} to be the inner products (α_i, α_j) of the simple roots of G and are the entries of the "symmetrized Cartan matrix" A of G. We can and will normalize the inner product so that all the diagonal entries are positive reals. These numbers have the following properties:

$$a_{ii} > 0$$
$$a_{ij} = a_{ji}$$
$$a_{ij} \leq 0 \text{ if } i \neq j$$
$$2a_{ij}/a_{ii} \in \mathbb{Z}.$$

The Cartan matrix is normalized so that it has integral entries and diagonal entries all equal to 2 but is not symmetric; for our purposes it is better to use matrices that are symmetric but do not in general have integral entries or diagonal coefficients equal to 2. It is easy to get from a symmetrized Cartan matrix to the Cartan matrix just by multiplying all the rows by suitable constants.

We can recover the Lie algebra G from the matrix A as the Lie algebra generated by an $sl_2 = \langle e_i, f_i, h_i \rangle$ for each simple root α_i, subject to the following relations (due to Serre and Harish-Chandra) depending on the numbers a_{ij}:

$$[e_i, f_i] = h_i$$
$$[e_i, f_j] = 0 \text{ for } i \neq j$$
$$[h_i, e_j] = a_{ij}e_j$$
$$[h_i, f_j] = -a_{ij}f_j$$
$$Ad(e_i)^{1-2a_{ij}/a_{ii}}e_j = 0$$
$$Ad(f_i)^{1-2a_{ij}/a_{ii}}f_j = 0$$

Kac and Moody noticed that we can define a Lie algebra in the same way for any matrix A satisfying the conditions above; these are the symmetrizable Kac-Moody algebras. (They also defined Lie algebras for non symmetrizable Cartan matrices, which we will not use.) Kac-Moody algebras have many of the properties of finite dimensional simple Lie algebras: we can define roots, Weyl chambers, Weyl groups, Cartan subalgebras, Verma modules, and so on by copying the usual definitions for finite dimensional Lie algebras. There is a Weyl-Kac character formula for the characters of some simple quotients of Verma modules, which for finite dimensional Lie algebras is just the usual Weyl character formula for finite dimensional representations. The only case of this we will use is the Weyl-Kac denominator formula, which is the Weyl-Kac character formula for the trivial one dimensional module of character 1, which says that

$$\sum_{w\in W} \det(w)w(e^\rho) = e^\rho \prod_{\alpha>0}(1-e^\alpha)^{mult(\alpha)}$$

where ρ is a special vector called the Weyl vector, the product is over all positive roots α, and *mult* is the multiplicity of a root, in other words the dimension of the corresponding root space.

Kac showed that we recover the Macdonald identities if we apply this denominator formula to certain special Kac-Moody algebras called affine Kac-Moody algebras, which are roughly central extensions of the Lie algebras $G[t, 1/t]$ of Laurent polynomials with coefficients in a finite dimensional Lie algebra G or twisted versions of these. For example the denominator formula of the Lie algebra $sl_2[t, 1/t]$ is the Jacobi triple product identity

$$\sum_{n\in\mathbb{Z}}(-1)^n q^{n^2} z^n = \prod_{n>0}(1-q^{2n})(1-q^{2n-1}z)(1-q^{2n-1}/z).$$

We would like to find some generalizations of the Macdonald identities corresponding to some Kac-Moody algebras other than the affine ones. To do this we need to find some Kac-Moody algebras for which both the root multiplicities $mult(\alpha)$ and the simple roots are known explicitly in some easy form. Knowing the simple roots is equivalent to knowing the Weyl group and hence the sum in the Weyl-Kac denominator formula, and knowing the root multiplicities is equivalent to knowing the product in the denominator formula. Unfortunately there are no known examples of Kac-Moody algebras (other than sums of finite dimensional and affine ones) for which both the simple roots and root multiplicities are known explicitly. It is of course always possible to calculate the multiplicity of any given root of any Lie algebra if we know the simple roots by using the denominator formula. This can be used to give either a recursive formula for the root multiplicities (due to Peterson) or a large and complicated alternating sum for the multiplicities. Unfortunately neither of these seems to give a satisfactory simple and explicit formula for the root multiplicities of any non affine Kac-Moody algebra. There have been several numerical calculations of the multiplicities of some of the easier Kac-Moody algebras, and the impression one gets from looking at these tables is that the root multiplicities look rather complicated and random.

3 Vertex algebras

As motivation for the definition of vertex algebras we first recall a short construction for the finite dimensional simple complex Lie algebras from their root lattice. We will just do the cases of the Lie algebras A_n, D_n, and E_6, E_7, E_8 (the others can be obtained as fixed points of these Lie algebras under diagram automorphisms). Suppose that L is the root lattice of G, so that the roots of G are exactly the norm 2 vectors of L. We construct a central extension $\hat{L}$ of L by a group of order 2 generated by an element ζ with the property that $e^a e^b = \zeta^{(a,b)} e^b e^a$ if e^a, e^b are lifts to $\hat{L}$ of $a, b \in L$. This central extension is unique up to (nonunique) isomorphism. We define G to be the Z-module

$$L \oplus \sum_{\alpha^2=2} e^\alpha$$

where the sum is over a set of lifts of the norm 2 vectors of L, and where we identify ζe^a with $-e^a$. We define the Lie bracket on G by

$$\begin{aligned} &[a, b] = 0 \text{ for } a, b \in L \\ &[a, e^b] = -[e^b, a] = (a, b)e^b \text{ if } a, b \in L, (b, b) = 2 \\ &[e^a, e^b] = a \text{ if } a = -b, e^a e^b \text{ if } (a, b) = -1, \end{aligned}$$

and 0 otherwise.

Then it is not hard to check that this bracket is antisymmetric and satisfies the Jacobi identity, so it defines an integral form of the complex Lie algebras G. (Notice that if we did not first take a central extension of L the product would not be antisymmetric. In fact there is no canonical way to construct G from L because the automorphism group of L, which is more or less the Weyl group of G, is not usually a subgroup of the automorphism group of G but is only a subquotient.)

This construction gives a completely explicit basis for the Lie algebra G. We would like to do something similar for all Kac-Moody algebras. Unfortunately the construction above breaks down as soon as the lattice L is indefinite and there are norm 2 vectors a, b with $(a, b) \leq -2$ but $a \neq b$. The vertex algebra of a lattice will provide a generalization of the construction above for all lattices.

For simplicity we will only construct the vertex algebra of a lattice L in which all inner products are even; in general if some inner products are odd it is necessary to first replace L by a central extension $\hat{L}$ as above. We will also only do the construction over the complex numbers, although with slightly more effort it can be done over the integers. We define the underlying space of the vertex algebra V of L to be the universal commutative ring with derivation D generated by the complex group ring $\mathbb{C}(L)$ of L. It is not hard to work out the structure of the ring V: it is just the tensor product for the group ring of L with the symmetric algebra of the sum of a countable infinite number of copies L_i of L. It is best to think of V as a commutative ring together with an action of the additive formal group $\mathbb{C}$ of the complex numbers, the action of the formal group being given by the derivation D (which spans the Lie algebra of this formal group).

A vertex operator is just a formal series $v(z) = \sum_{n\in\mathbb{Z}} v_{-n-1}z^n$ where the v_n's are $\mathbb{C}$-linear operators from V to V such that for any $w \in V$ the elements $v_n w$ vanish for n sufficiently large. We can think of $v(z)$ as being a sort of formal operator valued meromorphic function of z, with v_n given formally as the residue of $v(z)z^n dz$ at $z = 0$. We will now define some vertex operators on V. For each $a \in L$ we define the vertex operator a^+ by $a^+(z) = \sum_{n\geq 0} D^n(e^a)z^n/n!$ (where the operator $D^n(e^a)$ is the operator of multiplication by the element $D^n(e^a)$). We define another vertex operator a^- for $a \in L$ by saying that $a^-(z)$ is the derivation from V to $V[z, 1/z]$ taking e^b to $z^{(a,b)}e^b$. Then we can check that all the vertex operators of the form a^+ commute with each other, all the operators $a^-(z)$ commute with each other. Finally we define the vertex operators $a(z)$ by $a(z) = a^+(z)a^-(z)$ for $a \in L$, so that the operators $a(y)$ and $b(z)$ formally commute with each other for any $a, b \in L$. in the sense that

$$(y-z)^N\left(a(y)b(z) - b(z)a(y)\right) = 0$$

for N sufficiently large depending on a and b. Notice that the coefficient of $y^m z^n$ in $a(y)b(z)$ is not usually the same as the coefficient of $y^m z^n$ in $b(z)a(y)$; it is only when both sides are multiplied by a power of $y - z$ that the coefficients become equal.

We can now define a vertex algebra (roughly) as a vector space V such that for each element v of V we are given a vertex operator $v(z)$, and these vertex operators all formally commute with each other. We also require that V should have an identity element 1 such that $v(0)1 = v$ and $1(z)v = v$. The best way to think of a vertex algebra is as a sort of commutative ring with a formal action of the group $\mathbb{C}$, where the action of $z \in \mathbb{C}$ on v is $v(z)$, and $v(z)w$ is the ring product of $v(z)$ and w. If the vertex operator $v(z)$ is "holomorphic" for all v, which means that $v_n = 0$ for $n \geq 0$, then a vertex algebra is just a commutative ring with derivation and the ring product is defined by $vw = v_{-1}w$. In general we can think of V as behaving like a commutative ring whose multiplication is not defined everywhere because it has poles; if we perturb one of the elements v, w by an element z of the group $\mathbb{C}$ then we get a meromorphic function $v(z)w$ of z (behaving like the product of v acted on by z and of w) which will in general have a pole at $z = 0$.

Suppose we have a vector space V acted on by a set of commuting operators, such that V is generated as a vector space by the images of an element $1 \in V$ under the ring R generated by these operators. Then it is easy to see that the map from R to V taking r to $r(1)$ is an isomorphism of vector spaces and hence gives V the structure of a commutative ring with unit element 1. There is a similar theorem for vertex operators: if V is a vector space acted on by an operator D with a compatible set of commuting vertex operators acting on it, such that V is generated from an element $1 \in V$ with $D(1) = 0$ by the components of these vertex operators, then we can make V into a vertex algebra. If we apply this to the space V above constructed from a lattice and to the commuting set of operators $a(z)$ for $a \in L$ we see that V can be given a vertex algebra structure. If the inner product on L is identically zero then this vertex algebra structure on V is essentially the same as the commutative ring structure defined above; in general the vertex algebra structure defined by some other inner product on V can be thought of as a sort of "meromorphic deformation" of this ring structure.

If u, v, and w are elements of a vertex algebra then $u(x)v(y)w = v(y)u(x)w$ is a meromorphic function of x and y with poles only at $x = 0$, $y = 0$, and $x = y$. This function is also equal to $(u(x-y)v)(y)w$, which is easy to understand if one interprets $u(x)$ as the action of the group element x on the element u. (This is easier to see if we denote the action of a group element x on a ring element u by u^x, when it becomes $u^x v^y w = (u^{x/y}v)^y w$.) If $f(x)$ is any function of x with poles only at $x = 0$ or $x = y$ then Cauchy's formula shows that the residue at 0 plus the residue at y is the integral around a large circle of $f(x)dx/2\pi i$. If we apply this to $f(x) = u(x)v(y)w$ and then take residues at $y = 0$, we find that

$$\int_y \int_x (u(x-y)v)(y)w\,dx\,dy = \int_y \int_x u(x)v(y)w\,dx\,dy - \int_y \int_x v(y)u(x)w\,dx\,dy$$

where the paths of integration of x are a small circle around y, a large circle around both y and 0, and a small circle around 0 in the 3 integrals, or in other words

$$(u_0 v)_0 w = u_0(v_0 w) - v_0(u_0 w).$$

If we first multiply the integrands by $(x-y)^q x^m y^n$ we get more complicated identities involving the operators $u_i v$ and some binomial coefficients, which are (non trivially) equivalent to the identities first used to define vertex algebras in [B86].

4 The no-ghost theorem and I. Frenkel's upper bound

Recall that the Virasoro algebra is a central extension of the Lie algebra of polynomial complex vector fields on the circle, which is spanned by elements L_i, $i \in \mathbb{Z}$, with $[L_i, L_j] = (i-j)L_{i+j}$. The vertex algebra of a nonsingular lattice L has a natural action of the Virasoro algebra on it and we define the physical subspace M of $V(L)$ to be the quotient of P^1/DP^0 by the kernel of a certain bilinear form, where P^i is the space of (lowest weight) vectors with $L_0(v) = iv$, $L_i(v) = 0$, $i > 0$. (The motivation for this comes from string theory, where this space is roughly the space of physical states of a chiral string moving on the torus $L \otimes \mathbb{R}/L$.) The vertex algebra $V(L)$ also has a natural L-grading induced by the obvious L grading on the group ring of L. If L is 26 dimensional and Lorentzian then the no-ghost theorem states (among other things) that the piece of degree $\alpha \in L, \alpha \neq 0$ has dimension $p_{24}(1-\alpha^2/2)$, where $p_{24}(n)$ is the number of partitions of n into parts of 24 colors. (In spite of several statements to the contrary in the mathematical literature, the original proof of the no-ghost theorem by Goddard and Thorn is mathematically rigorous.)

I. Frenkel showed [F] that if L is the lattice $II_{25,1}$ then the Kac-Moody algebra M_{KM} (with Dynkin diagram given by the Dynkin diagram of $II_{25,1}$) can be embedded as a subspace of the space of physical states of $V(II_{25,1})$, and in particular this gives $p_{24}(1-\alpha^2/2)$ as an upper bound on the multiplicities of the root spaces of this Lie algebra. (This work of Frenkel's was the main motivation for the definition of vertex algebras.)

Frenkel's method gives the same upper bound for the multiplicities of the roots of any Kac-Moody algebra of rank 26 all of whose roots have norm 2. For Kac-Moody algebras of rank k not equal to 26 it gives the weaker bound $p_{k-1}(1-\alpha^2/2) - p_{k-1}(-\alpha^2/2)$ which is slightly larger than $p_{k-2}(1-\alpha^2/2)$. There are examples of Lie algebras of ranks not equal to 26, (such as E_{10}, due to Kac and Wakimoto), some of whose root multiplicities are strictly larger than $p_{k-2}(1-\alpha^2/2)$, so the upper bound given by the no ghost theorem in rank 26 cannot be generalized in the obvious way to all ranks.

We can next ask how good Frenkel's upper bound $p_{24}(1-\alpha^2/2)$ for the root multiplicities of M_{KM} is. If we calculate the multiplicities of the roots of M_{KM} (for example by using the Peterson recursion formula) we find the following results. All norm 2 vectors have multiplicity 1, equal to Frenkel's upper bound $p_{24}(1-2/2)$. For norm 0 vectors there are 24 orbits of primitive norm 0 vectors corresponding to the 24 Niemeier lattices (24 dimensional even unimodular positive definite lattices). The correspondence is given as follows: if w is a nonzero norm 0 vector in $II_{25,1}$ then the quotient $w^\perp/w$ of the orthogonal complement $w^\perp$ of w by the space generated by w is a Niemeier lattice, and conversely if N is a Niemeier lattice then $N \oplus II_{1,1}$ is a 26 dimensional even Lorentzian lattice and is therefore isomorphic to $II_{25,1}$. A norm 0 vector in the $II_{1,1}$ gives a norm 0 vector w in $II_{25,1}$ with $w^\perp/w = N$. If z is any nonzero norm 0 vector in $II_{25,1}$ then it is not hard to check using the theory of affine Lie algebras that the multiplicity of the root nz is equal to the rank of the corresponding Niemeier lattice, which is 0 for the Leech lattice and 24 for any other Niemeier lattice. So the multiplicity of the vector nz is equal to Frenkel's upper bound $p_{24}(1-0^2/2) = 24$ except when z is a norm 0 vector corresponding to the Leech lattice. For norm -2 vectors the calculations take much more effort. There are 121 orbits of norm -2 vectors. One of the orbits has multiplicity 0, one has multiplicity 276, and the other 119 orbits all have multiplicity $324 = p_{24}(1-(-2)/2)$ equal to Frenkel's upper bound. Similarly there are 665 orbits of norm -4 vectors, all but 3 of which turn out to have multiplicity equal to $3200 = p_{24}(1-(-4)/2)$. Furthermore when calculating these multiplicities using (say) the Peterson recursion formula, it is apparent that the "deficiencies" in the multiplicities of vectors of negative norm are "caused" by the fact that the multiplicities of the norm 0 vectors corresponding to the Leech lattice are 0 rather than 24. This suggests that if we somehow added a 24 dimensional root space for each multiple of the norm 0 vector ρ then we would get a Lie algebra whose root multiplicities were exactly equal to Frenkel's upper bound. We already have a space M containing M_{KM} given by the space of all physical states of a chiral string in 26 dimensions. The remarks and calculations above strongly suggest that M itself should be a Lie algebra with simple roots given by the simple roots of M_{KM} together with multiples of ρ.

It is easy to construct a Lie algebra product on M using the theory of vertex algebras (which is not surprising given that vertex algebras were invented partly to construct such a Lie bracket). At the end of section 3 we saw that the vertex algebra product $u_0v = u_0(v)$ satisfies the identity

$$(u_0v)_0w = u_0(v_0w) - v_0(u_0w)$$

which is one version of the Jacobi identity, but it is not antisymmetric. However the sum $u_0v + v_0u$ does at least lie in the image of the operator $D = L_{-1}$, and using this it is easy to check that defining $[u, v]$ to be u_0v defines a Lie bracket on the quotient V/DV for any vertex algebra V (where DV is the image of V under the derivation D). The space M is a subquotient of the Lie algebra V/DV where V is the vertex algebra of the lattice $II_{25,1}$, so this defines a Lie algebra structure on M. (In fact M is even a subalgebra of V/DV, but this is harder to prove and depends heavily on the structure of M described later.)

There is another way to construct the Lie algebra M using semi-infinite cohomology. It follows from [FGZ] that the vector space M can be identified with a certain semi-infinite cohomology group, which therefore has a natural Lie algebra structure. Lian and Zuckerman study this in [LZ] and reconstruct this Lie algebra from an algebraic structure called a Gerstenhaber algebra on the full semi-infinite cohomology. In the case of the fake monster Lie algebra M we do not really get anything new because the semi-infinite cohomology with its algebraic structure can be reconstructed from the Lie algebra M with its bilinear form, but in more general cases this is no longer true, and semi-infinite cohomology is probably the correct way to construct Lie algebras.

We would now like to work out the structure of M. As M is close to being a Kac-Moody algebra this suggests that we should try to "force" M to be a Kac-Moody algebra and find out what the obstruction to this is. We summarize the facts we know about M: it is graded by $II_{25,1}$ with root multiplicities given by $p_{24}(1-\alpha^2/2)$, it has an invariant bilinear form $(,)$ induced by the form on the vertex algebra V, it has an involution ω induced by the automorphism -1 of $II_{25,1}$ (or more precisely by a lift of this to a double cover of $II_{25,1}$), and the contravariant form $(a,b)_0=(a,\omega(b))$ is positive definite on the root spaces of all nonzero roots. (The form $(,)_0$ is not positive definite on the zero weight space M_0; this weight space is 26 dimensional and the form $(,)_0$ on it is Lorentzian. So M is an infinite dimensional Lorentzian space under $(,)_0$, in other words, it has a negative norm vector whose orthogonal complement is positive definite.) We try to make M into a Kac-Moody algebra by finding the elements e_i and then defining the remaining generators f_i and h_i by $f_i=\omega(e_i), h_i=[e_i,f_i]$. Any Kac-Moody algebra G can be written as a direct sum of subspaces $E\oplus H\oplus F$ where H is the Cartan subalgebra and E and F are the sums of the root spaces of the positive and negative roots. (Of course G is not the Lie algebra sum of the subalgebras E, F, and H, but only the vector space sum.) If we choose a positive integer n_i for each simple root α_i then we can $\mathbb{Z}$-grade G by giving e_i degree n_i, f_i degree $-n_i$, and h_i degree 0. Then $E=\oplus_{n>0}E_n$ is the sum of the positive degree subspaces E_n of G. The elements e_i are then a minimal set of generators of the space E, in the sense that the elements e_i of any fixed degree n are a basis for the quotient of E_n by the elements in the subalgebra generated by the E_k's for $k<n$. This suggests that we can

recursively construct the elements e_i.

We carry out this process for the Lie algebra M. We grade it by choosing any negative norm vector v which has nonzero inner product with all norm 2 vectors, and defining the degree of M_r to be (r, v), so that we get a subalgebra $E = \oplus_{n>0} E_n$ of the positive root spaces of M. We recursively construct the elements e_i in E_n as a basis for the space of elements of E_n that are orthogonal to the subalgebra generated by all the e_i's we have previously constructed in E_k for $k < n$. Then all the e_i's we construct in this way generate E, essentially because the form $(,)_0$ is positive definite on E_n so that any space E_n is the sum of any subspace and its orthogonal complement. We can also arrange that the e_k's we construct are orthonormal and are eigenvectors of $H = M_0$. Finally we define the generators f_i and h_i in terms of e_i by $f_i = \omega(e_i), h_i = [e_i f_i]$ as above, and define the elements a_{ij} of a matrix A by $a_{ij} = (h_i, h_j)$. Now we ask whether the elements a_{ij} and the generators e_i, f_i, and h_i satisfy the conditions for Kac-Moody algebras in section 2. It turns out that they do not quite: instead they satisfy the slightly weaker conditions

$$a_{ij} = a_{ji}$$

$$a_{ij} \leq 0 \text{ if } i \neq j$$

$$2a_{ij}/a_{ii} \in \mathbb{Z} \text{ if } a_{ii} > 0$$

$$[e_i, f_i] = h_i$$

$$[e_i, f_j] = 0 \text{ for } i \neq j$$

$$[h_i, e_j] = a_{ij} e_j$$

$$[h_i, f_j] = -a_{ij} f_j$$

$$Ad(e_i)^{1-2a_{ij}/a_{ii}} e_j = 0 \text{ if } a_{ii} > 0,\ [e_i, e_j] = 0 \text{ if } a_{ij} = 0.$$

$$Ad(f_i)^{1-2a_{ij}/a_{ii}} f_j = 0 \text{ if } a_{ii} > 0,\ [f_i, f_j] = 0 \text{ if } a_{ij} = 0.$$

If all the diagonal entries a_{ii} are positive, in other words if all the simple roots are real (positive norm), then these conditions are exactly those defining symmetrizable Kac-Moody algebras. We will call any Lie algebra generated by relations satisfying the conditions above a generalized Kac-Moody algebra. (In fact generalized Kac-Moody algebras are slightly more general than this, because we also allow a few extra operations, such as adding outer derivations, quotienting by subspaces of H in the center of G, and taking central extensions.)

As the name implies, the theory of generalized Kac-Moody algebras is similar to that of Kac-Moody algebras (and finite dimensional simple Lie algebras), with some minor changes. The main difference is that we may have imaginary (norm ≤ 0) simple roots. These contribute some extra terms to the sum in the Weyl-Kac character formula. In particular the denominator formula now looks like

$$\sum_{w\in W}\det(w)\sum_{S}(-1)^{|S|}w(e^{\rho+\sum S}) = e^{\rho}\prod_{\alpha>0}(1-e^{\alpha})^{mult(\alpha)}.$$

The inner sum is over sets S of pairwise orthogonal imaginary simple roots, of cardinality $|S|$ and sum σS. Notice that knowing the sum on the left is essentially equivalent to knowing all the simple roots, because the real simple roots determine the Weyl group W in the outer sum, and the imaginary simple roots determine the inner sum. There is an explicit example of such a denominator formula at the end of this section.

Roughly speaking the imaginary simple roots do not really add anything very interesting to G; the complexity of G is closely related to the complexity of the Weyl group W of G, which only depends on the real simple roots of W. The imaginary simple roots only contribute some extra flab, which usually looks a bit like free or free abelian Lie algebras [J]. The main advantage of generalized Kac-Moody algebras is that, as we have seen above for M, it is sometimes possible to prove that some naturally occurring Lie algebra is a generalized Kac-Moody algebra using only general properties of M (in particular the existence of an "almost positive definite" contravariant bilinear form $(,)_0$). Most "natural" examples of Kac-Moody algebras other than affine ones (such as the Kac-Moody algebra M_{KM} with Dynkin diagram the Leech lattice, or the Frenkel-Feingold algebra [FF]) seem to be the subalgebra of some more natural generalized Kac-Moody algebra (such as the fake monster Lie algebra M) generated by some of the real simple root spaces. See the end of this section for an example.

So far we have seen that M is a generalized Kac-Moody algebra with known root multiplicities, so the next obvious thing to do is to try to work out the simple roots of M. The fact that we will be able to do this for M may be misleading: there are very few known examples of generalized Kac-Moody algebras for which the simple roots and root multiplicities are both known. For example it is possible to construct a generalized Kac-Moody algebra similar to M for any Lorentzian lattice L, but if L is any Lorentzian lattice other than $II_{25,1}$ then the simple roots of the generalized Kac-Moody algebra of L seem to be too messy to describe explicitly. Most of the several hundred known examples of generalized Kac-Moody superalgebras whose simple roots and root multiplicities are both known explicitly can be obtained by "twisting" the denominator formula for the fake monster Lie algebra in some way.

There are some obvious simple roots of M: the simple roots $(\lambda, 1, \lambda^2/2-1)$ of M_{KM} are the real simple roots of M, and the positive multiples of the norm 0 Weyl vector $\rho = (0,0,1)$ of M are simple roots of multiplicity 24. The calculations of root multiplicities above turn out to be equivalent to saying that M has no simple roots of norm -2 or -4, and suggest that M has no simple roots other than the ones of norm 2 or 0 above. This can be proved roughly as follows. We can get information about the imaginary simple roots using the fact that they appear in the sum defining the denominator function. The denominator function is not easy to describe completely, so the the key step is to look at the denominator function restricted to the 2 dimensional space of vectors of the form $(0, \sigma, \tau) \in II_{25,1} \otimes \mathbb{C}$ with $\Im(\sigma) > 0$, $\Im(\tau) > 0$. The restriction to this space of the infinite product defining the denominator function is

$$p^{-1} \prod_{m>0, n\in\mathbb{Z}} (1-p^m q^n)^{c'(mn)}$$

where $p = e^{2\pi i\sigma}$, $q = e^{2\pi i\tau}$, and the numbers $c'(mn)$ are defined by

$$\sum_{n\in\mathbb{Z}} c'(n)q^n = \Theta_\Lambda(\tau)/\Delta(\tau) = j(\tau) - 720 = q^{-1} + 24 + 196884q + 21493760q^2 + \cdots$$

The function $1/\Delta(\tau)$ has the numbers $p_{24}(n+1)$ as the coefficient of q^n, and the function Θ_Λ is the theta function of the Leech lattice, which appears because the Leech lattice is the kernel of the projection of $II_{25,1}$ to the 2 dimensional lattice $II_{1,1}$. Their product is, up to a constant, the elliptic modular function $j(\tau)$. This infinite product can be evaluated explicitly by rewriting it as

$$p^{-1} \exp\Big(\sum_{m>0} p^m T_m(j(\tau) - 720))\Big)$$

where the T_m's are Hecke operators. The point is that if T_m is applied to a modular function such as j then the result is still a modular function (and in fact a polynomial in j). In particular with respect to the variable τ the infinite product behaves like a modular form of weight 12. It is also antisymmetric in σ and τ, and these two conditions (plus some technical conditions about growth at infinity) are sufficient to characterize it up to multiplication by a constant. The result is that

$$p^{-1} \prod_{m>0, n\in\mathbb{Z}} (1-p^m q^n)^{c'(mn)} = \Delta(\sigma)\Delta(\tau)(j(\sigma) - j(\tau)).$$

This identity turns out to mean that in some sense the multiplicities of the root spaces of imaginary simple roots of $II_{25,1}$ with the same image in $II_{1,1}$ have an "average" value of 0 (in some rather weird sense of the word average). As the multiplicity of a simple root is the dimension of the simple root space and hence nonnegative, the only way in which the simple root multiplicities can have an average value of 0 is if they are all 0. In other words there are no simple roots of negative norm. For more precise details of this argument see [B90].

The denominator formula for M can now be written out explicitly as we know both the simple roots and the root multiplicities, and it is ([B90])

$$e^{\rho}\prod_{r>0}(1-e^{r})^{p_{24}(1-r^2/2)} = \sum_{w\in W n\in Z} \det(w)\tau(n)e^{w(n\rho)}$$

where $\tau(n)$ is the Ramanujan tau function defined by

$$\sum_{n\in\mathbb{Z}}\tau(n)q^n = q\prod_{n>0}(1-q^n)^{24} = q-24q^2+\cdots .$$

The terms with $n=1$ on the right hand side are exactly what one would get for the denominator formula of the Kac-Moody algebra M_{KM}; the terms with $n>1$ are the extra correction terms coming from the imaginary simple roots (= positive multiples of ρ) of M.

We have seen that the Kac-Moody algebra with Dynkin diagram given by the Leech lattice is best thought of as a subalgebra of a larger generalized Kac-Moody algebra whose root multiplicities are known explicitly. A similar phenomenon also happens for some other non affine Kac-Moody algebras; they are best thought of as large subalgebras of generalized Kac-Moody algebras. We will illustrate this by discussing the case of Frenkel and Feingold's algebra [FF] with Cartan matrix

$$\begin{pmatrix} 2 & -1 & 0 \\ -1 & 2 & -2 \\ 0 & -2 & 2 \end{pmatrix}$$

which can be thought of as the affine A_1 Dynkin diagram with an extra node attached. Frenkel and Feingold calculated many of the root multiplicities of roots λ, and observed that many of them were given by values $1,1,2,3,5,7,11,\ldots$ of the partition function $p(1-\lambda^2/2)$. A suitable generalized Kac-Moody algebra containing it was described by Niemann in his thesis [N]. The root lattice of this Lie algebra is $K\oplus II_{1,1}$ where K is generated by x and y with $(x,x)=4$, $(y,y)=6$, $(x,y)=-1$. This lattice has determinant 23 and no roots, and corresponds to a non principal ideal in the imaginary quadratic field of discriminant -23. We define $p_\sigma(n)$ by

$$\sum_n p_\sigma(n)q^n = \frac{1}{\eta(\tau)\eta(23\tau)} = q^{-1} + 1 + 2q + 3q^2 + 5q^3 + 7q^4 + 11q^5 + \cdots$$

Notice that the first 23 values of $p_\sigma(n)$ are the same as the values of the partition function $p(n+1)$. Niemann showed that there is a generalized Kac-Moody algebra with root lattice K, whose root multiplicities are given by

$$mult(\lambda) = p_\sigma(-\lambda^2/2)$$

if $\lambda \in K, \lambda \notin 23K'$ and by

$$mult(\lambda) = p_\sigma(-\lambda^2/2) + p_\sigma(-\lambda^2/46)$$

if $\lambda \in 23K'$. Moreover this Lie algebra has a norm 0 Weyl vector and its simple roots can be described explicitly in a way similar to those of the fake monster Lie algebra. In particular the norm 2 simple roots correspond to points of the lattice K, and there are also some norm 46 simple roots corresponding to two cosets of K, and some norm 0 simple roots of multiplicity 1 or 2 corresponding to positive multiplicities of the Weyl vector. If we take the norm 2 simple roots corresponding to 0 and the 2 basis vectors of K we find they have the same Dynkin diagram as the Feingold-Frenkel Lie algebra, so in particular the Feingold-Frenkel Lie algebra is a subalgebra and its root multiplicities are bounded by the multiplicities given above. It is also easy to check that in some sense the Feingold-Frenkel Lie algebra accounts for all the "small" roots, so the first few root multiplicities are given exactly by $p_\sigma(-\lambda^2/2)$, which are values of the partition function, and this explains the observation that the Feingold-Frenkel Lie algebra's root multiplicities are often given by values of the partition function. So this answers Frenkel and Feingold's question about whether it is possible to write down an explicit denominator formula for their Lie algebra, as long as we are willing to modify their question a bit and look at a slightly bigger Lie algebra.

Niemann also proved similar results for some other Kac-Moody algebras obtained by adding an extra node to an affine Dynkin diagram.

5 Relations with moonshine

The identities above are used in the proof of the moonshine conjectures, which state that the monster sporadic group has an infinite dimensional graded representation $V = \oplus V_n$ such that the traces of elements of the monster on this representation are given by certain Hauptmoduls. In particular the dimensions of the graded pieces V_n are given by the coefficients of q^{n-1} of the elliptic modular function $j(\tau) - 744$. As there are already several survey articles ([B94], [LZ], [J], [G], [Y]) about the proof of the moonshine conjectures we will only briefly discuss its relation with the fake monster Lie algebra.

A candidate for the representation V was constructed by Frenkel, Lepowsky and Meurman [FLM] and has the structure of a vertex algebra [B86]. This vertex algebra is a twisted version of the vertex algebra of the Leech lattice. If we look at the restriction of the denominator function of the fake monster Lie algebra to the 2 dimensional subspace (see the end of section 4) we see that it says an infinite product is equal to a simple expression, so we can ask if this restriction is itself the denominator function of some generalized Kac-Moody algebra. This is not quite right, but if we divide both sides by $\Delta(\sigma)\Delta(\tau)$ we find the identity

$$p^{-1} \prod_{m>0, n\in Z} (1 - p^m q^n)^{c(mn)} = j(p) - j(q).$$

This is the denominator formula of a $II_{1,1}$-graded generalized Kac-Moody algebra whose piece of degree $(m,n) \neq (0,0)$ is equal to the coefficient $c(mn)$ of q^{mn} in $j(\tau) - 744 = \sum_n c(n)q^n = q^{-1} + 196884q + \cdots$. These numbers are exactly the same as the dimensions of the graded pieces of V, and using this as a hint it is easy to guess how to construct a generalized Kac-Moody algebra from V called the monster Lie algebra: we first tensor the vertex algebra V with the vertex algebra of $II_{1,1}$ to get a vertex algebra similar to that of the lattice $II_{25,1}$, and then we apply the same construction to $V \otimes V(II_{1,1})$ that we used to construct the fake monster Lie algebra from $V(II_{25,1})$. In other words the monster Lie algebra is just the space of physical states in $V \otimes V(II_{25,1})$. As V is a twisted version of $V(\Lambda)$, we see that the monster Lie algebra is a twisted version of the fake monster Lie algebra; more precisely, both Lie algebras have automorphisms of order 2 such that the fixed point subalgebras are isomorphic. The monster Lie algebra can also be constructed via semi infinite cohomology as in [FGZ, LZ].

The monster Lie algebra by construction has an action of the monster on it with the property that the pieces of degrees (a,b) and (c,d) are isomorphic as representations of the monster whenever $ab = cd \neq 0$. This can be used to prove Conway and Norton's moonshine conjectures; see [B92], [J], [G], [B94].

6 The denominator function

The sum in the denominator formula for the fake monster Lie algebra defines a function Φ on a subset of $II_{25,1} \otimes \mathbb{C}$. It is not hard to check that this sum converges whenever the imaginary part of the argument lies in the interior of the cone C. The fact that Φ can also be written as an infinite product suggests that we can find the zeros of Φ by looking at the zeros of the factors in the infinite product. This turns out to be incorrect because the infinite product does not converge everywhere and the function Φ has zeros outside the region of convergence. (This is a quite common phenomenon; for example, it also happens for the Riemann zeta function!) We can find one zero of Φ which cannot be seen in any of its factors as follows.

First suppose that we consider Φ just for purely imaginary values of its argument v. Then we can work out the exact region of convergence of the infinite product, because we know the asymptotic behavior of the coefficients $p_{24}(1+n)$ by the Hardy-Ramanujan theorem, and this region turns out to be exactly the region with $v^2 > 2$. (Notice the region is $v^2 > 2$ rather than $v^2 < -2$ because v is purely imaginary.) Next we can see that Φ vanishes whenever v is purely imaginary and $v^2 = 2$. To see this recall the well known lemma from complex analysis which says that if $f(z) = \sum a_n z^n$ is a power series with radius of convergence r and all the a_n's are non negative, then f has a singularity at $z = r$. All the coefficients of the series for $-\log(\Phi)$ are non negative (as can be seen by taking the log of the infinite product and using the fact that $p_{24}((1+n)$ is nonnegative), so by using a higher dimensional version of the lemma mentioned above we see that $\log(\Phi(v))$ has a singularity whenever v is purely imaginary and $v^2 = 2$. However Φ itself is holomorphic at these points as this is inside the region of convergence of the infinite sum defining Φ, so the only way that $\log(\Phi(v))$ can be singular is if Φ vanishes at these points. Finally as Φ is holomorphic it must vanish at all points of the divisor $v^2 = 2$ inside the region where it is defined. Notice that this divisor is not a zero of any factor of the infinite product for Φ and (hence) lies entirely outside the region of convergence of this infinite product.

The fact that Φ vanishes for $v^2 = 2$ suggests that perhaps Φ satisfies some sort functional equation forcing it to vanish at these points, perhaps something like $\Phi(2v/(v,v)) = f(v)\Phi(v)$ where f is some function not equal to 1 when $(v,v) = 2$. We can guess the exact form of f by looking at Φ restricted to the 2 dimensional space of vectors $(0, \sigma, \tau)$, where we evaluated Φ explicitly as $\Delta(\sigma)\Delta(\tau)(j(\sigma) - j(\tau))$. Using the functional equations $\Delta(-1/\tau) = \tau^{12}\Delta(\tau)$, $j(-1/\tau) = j(\tau)$, and the fact that if $v = (0, \sigma, \tau)$

then $v^2 = -2\sigma\tau$ we see that on this 2 dimensional subspace Φ satisfies the functional equation

$$\Phi(2v/(v,v)) = -((v,v)/2)^{12}\Phi(v).$$

Once we have guessed the correct form of the functional equation above it is surprisingly easy to prove. As before we restrict to purely imaginary values of v because if we prove it for these v it will follow for all v by analytic continuation. We first observe that both $\Phi(v)$ and $(v,v)^{-12}\Phi(2v/(v,v))$ are solutions to the wave equation. For Φ this follows from the fact that it is a sum of terms of the form $e^{2\pi i(z,v)}$ with $z^2 = 0$, which are all solutions of the wave equation. If Φ is any solution of the wave equation in $\mathbb{R}^{25,1}$ then $(v,v)^{1-\dim(\mathbb{R}^{25,1})/2}\Phi(2v/(v,v))$ is automatically also a solution by the standard behavior of the wave operator under the conformal transformation $v \mapsto 2v/(v,v)$. (More generally we get an action of not just isometries but also the whole conformal group on the space of solutions of the wave equation.) Secondly, the fact that Φ vanishes for $v^2 = 2$ easily implies that both Φ and $-((v,v)/2)^{-12}\Phi(2v/(v,v))$ have the same partial derivatives of order at most 1 on this surface. (It is obvious they both vanish there and have vanishing first partial derivatives tangent to this surface, so we only need to check the first derivatives normal to the surface which is not hard.) Hence Φ and $-((v,v)/2)^{-12}\Phi(2v/(v,v))$ are both analytic solutions to a second order differential equation and both have the same derivatives of order at most 1 along some (non characteristic) Cauchy surface, so by the Cauchy Kovalevsky theorem they are equal.

7 The automorphic form Φ

We have seen that Φ satisfies the functional equation

$$\Phi(2v/(v,v)) = -((v,v)/2)^{12}\Phi(v),$$

and it is easy to check that it also satisfies the functional equations $\Phi(v+\lambda) = \Phi(v)$ for $\lambda \in II_{25,1}$ and $\Phi(w(v)) = \det(w)\Phi(v)$ for $w \in Aut(II_{25,1})^+$. We can work out the group generated by these three sorts of transformations and it turns out to be isomorphic to the group $Aut(II_{26,2})^+$, which is a discrete subgroup of the group $O_{26,2}(\mathbb{R})$ of conformal transformations of (a conformal completion of) $\mathbb{R}^{25,1}$. The action of $O_{26,2}(\mathbb{R})^+$ on the domain of definition of Φ is given as follows. We can identify this domain of vectors $v \in II_{25,1} \otimes \mathbb{C}$ with $\Im(v) \in C$ with a subset of norm 0 vectors in the projective space $P((II_{25,1} \oplus II_{1,1}) \otimes \mathbb{C})$ by mapping v to the norm 0 vector $(v, 1, v^2/2) \in II_{25,1} \oplus II_{1,1} = II_{26,2}$. The image is the set of vectors in projective space represented by norm 0 vectors whose real and imaginary parts form a positively oriented orthonormal base for a negative definite subspace of $\mathbb{R}^{26,2}$. As $O_{26,2}(\mathbb{R})^+$ obviously acts naturally on this space we get an action on the domain of definition of Φ.

We can think of the domain of Φ as a generalization of the upper half plane, and $O_{26,2}(\mathbb{R})^+$ as a generalization of the group $SL_2(\mathbb{Z})$ acting on the upper half plane. The function Φ should then be thought of as a generalization of a modular form on the upper half plane; these generalizations are called automorphic forms. We can summarize much of what we have discussed so far by saying that the denominator function of the fake monster Lie algebra is an automorphic form of weight 12 for the discrete subgroup $Aut(II_{26,2})^+$ of $O_{26,2}(\mathbb{R})^+$.

8 The zeros of Φ

We saw above that the naive guess for the zeros of Φ was wrong: there are zeros which cannot be seen in the factors of Φ. However we can now find all the zeros using the fact that Φ is an automorphic form. We have seen that Φ has a zero along the divisor $v^2 = 2$, and obviously also has zeros along all conjugates of this divisor under the group $Aut(II_{26,2})^+$ because of the transformations of Φ under this group. If we identify the domain of Φ with a subset of complex projective space as above these conjugates are easy to visualize: the divisor $v^2 = 2$ is just the set of points $(v, 1, v^2/2)$ orthogonal to the norm 2 vector $(0, 1, -1) \in II_{26,2}$, and $Aut(II_{26,2})^+$ acts transitively on these norm 2 vectors, so the zeros of Φ conjugate to $v^2 = 2$ just correspond to pairs $\{r, -r\}$ of norm 2 vectors in $II_{26,2}$. These are exactly the zeros of Φ that are "forced" by its functional equations. Now we want to see that Φ has no other zeros. Recall that the zero $v^2 = 2$ of Φ was somehow very closely related to the asymptotic behavior of the function $p_{24}(1+n)$. The Hardy-Ramanujan-Rademacher circle method gives a much finer description of this asymptotic behavior as

$$p_{24}(1+n) = 2\pi n^{-13/2} \sum_{k>0} \frac{I_{13}(4\pi\sqrt{n}/k)}{k} \sum_{0 \le h,h' < k hh' \equiv -1 \bmod k} e^{2\pi i(nh+h')/k}$$

where $I_{13}(z) = -iJ_{13}(iz)$ is the modified Bessel function of order 13. (In particular $p_{24}(1+n)$ is asymptotic to $2^{-1/2}n^{-27/4}e^{4\pi\sqrt{n}}$ for large n.) The zero $v^2 = 2$ corresponds to the dominant term in the asymptotic expansion, and it turns out that all other zeros are related to smaller terms in this expansion. The reason for this is that all singularities of a periodic function, such as $\log(\Phi)$, are very closely related to the asymptotic behavior of its Fourier coefficients (and in the same way the local behavior of any function depends on the growth at infinity of its Fourier transform). It turns out to be rather hard to describe the zeros of Φ directly like this, although it would in principle be possible. However it is not hard to see using these ideas that all zeros of Φ must be divisors corresponding to some positive norm vectors in $II_{26,2}$, perhaps of norms greater than 2. But every such divisor is conjugate under the group $Aut(II_{26,2})$ to some divisor intersecting the region of convergence of the infinite product for Φ. As it is easy to see that none of the factors of Φ have any zeros corresponding to vectors of $II_{26,2}$ of norms greater than 2, this shows that the only zeros of Φ are the ones we have already found.

9 Siegel theta functions

The proof that Φ is an automorphic form and has zeros as above is very indirect, using the no-ghost theorem, vertex algebras, generalized Kac-Moody algebras, and so on. As the statement only involves automorphic forms, we can ask if it is possible to prove that Φ is an automorphic form using only automorphic form theory. This can be done roughly as follows. (We will only sketch this method briefly, because we will see later in this section that there is a better way of proving this.) First by extending the argument in the previous section, we can prove that Φ (defined as the infinite product) can be analytically continued from the region of convergence of the infinite product to the full domain of Φ. Secondly we can write Φ in the form $\exp(\sum_{m>0} T_m(f))$ where f is a certain Jacobi form with poles at cusps and the T_m's are Hecke operators acting on this Jacobi form. (The idea of considering $\sum_{m>0} T_m(f)$ for holomorphic Jacobi forms f was used by Maass in his work on the Saito-Kurokawa conjecture described in [EZ] and then extended by Gritsenko [Gr] to Jacobi forms on higher dimensional lattices.) Notice that this is a sort of blown up version of the formula we found for the restriction of Φ to a 2 dimensional subspace in section 4. This second expression for Φ easily implies that it transforms nicely under a certain subgroup $SL_2(\mathbb{Z})$ of $Aut(II_{26,2})^+$. By looking at the Fourier coefficients of Φ it is easy to check that it also transforms nicely under $Aut(II_{25,1})^+$. As this group together with $SL_2(\mathbb{Z})$ generate $Aut(II_{26,2})^+$ we see that Φ transforms correctly under all elements of $Aut(II_{26,2})^+$ and hence shows that Φ is an automorphic form.

The advantage of the proof sketched above is that it does not really depend on all the special properties of the Leech lattice, and can be extended to more general lattices than $II_{25,1}$ and more general exponents that $p_{24}(1+n)$. If we carry out this extension for even unimodular lattices we find:

Theorem [B95]. *Suppose that L is the even unimodular lattice $II_{s+2,2}$. Suppose that $f(\tau)=\sum_n c(n)q^n$ is a meromorphic modular form of weight $-s/2$ for $SL_2(\mathbb{Z})$ with integer coefficients, with poles only at cusps, and with $24|c(0)$ if $s=0$. There is a unique vector $\rho\in L$ such that*

$$\Phi(v)=e^{-2\pi i(\rho,v)}\prod_{r>0}(1-e^{-2\pi i(r,v)})^{c(-(r,r)/2)}$$

is a meromorphic automorphic form of weight $c(0)/2$ for $Aut(II_{s+2,2})^+$.

It is also possible to describe the zeros and poles of Φ explicitly. The denominator function for the fake monster Lie algebra is the special case of the theorem above with $s=24$, $f(\tau)=1/\Delta(\tau)$.

The proof sketched above is not very enlightening, as it is essentially a brute force check that Φ transforms correctly under a set of generators of $Aut(II_{s+2,2})^+$. It is possible to extend this a bit to non unimodular lattices, but it becomes increasingly complicated as the level and determinant increase. Gritsenko and Nikulin [GN] have worked out a some higher level examples explicitly using the method above. In a recent preprint [HM], Harvey and Moore have sketched a conceptual proof of the fact that Φ is an automorphic form, and also given a better explanation of why its only zeros are as described above, using Siegel theta functions of indefinite lattices (which seem to have been independently rediscovered by physicists working in string theory). Their method generalizes much better to higher levels; see [B].

The Siegel theta function $\Theta_L(\tau_1,\tau_2,V)$ of a lattice $L=II_{r,s}$ (which we will assume is even and unimodular for simplicity) is defined as

$$\Theta_L(\tau_1,\tau_2,V)=\sum_{\mathfrak{l}\in L}\exp(2\pi i\tau_1\mathfrak{l}_1^2/2+2\pi i\tau_2\mathfrak{l}_2^2/2)$$

where $\Im(\tau_1)>0$, $\Im(\tau_2)<0$, V is a negative definite subspace of $L\otimes\mathbb{R}$ of maximal dimension n, and $\mathfrak{l}_1$ and $\mathfrak{l}_2$ are the projections of $\mathfrak{l}$ into $V^\perp$ and V. If L is positive definite this is just the usual theta function of L, as it does not depend on τ_2 or V, and for indefinite lattice the series still converges absolutely because of the condition that $\Im(\tau)_2 < 0$. It satisfies the functional equations

$$\Theta_L\left(a\tau_1 + \frac{b}{c}\tau_1 + d, a\tau_2 + \frac{b}{c}\tau_2 + d, V\right)$$
$$= ((c\tau_1 + d)/i)^{r/2}((c\tau_2 + d)i)^{s/2}\Theta_L(\tau_1, \tau_2, V),$$

and is obviously invariant under the natural action of $Aut(II_{r,s})$ on V.

Now suppose that f is any linear functional from functions of τ_1 and τ_2 to $\mathbb{C}$. Then $f(\Theta_L)(V)$ is a function of V which is automatically invariant under $Aut(II_{r,s})$.

Harvey and Moore make the following choices for L and f. They choose L to be a lattice $II_{s+2,2}$. The set of subspaces V can be identified with the hermitian symmetric space of L as follows: the norm 0 vector $x+iy \in L\otimes\mathbb{C}$ representing a point of the projective space $P(L\otimes\mathbb{C})$ corresponds to the negative definite space spanned by x and y. The linear functional they use is (more or less) given by

$$f(\Theta_L(\tau_1, \tau_2, V)) = \int_D \Theta_L(\tau, \bar\tau, v)g(\tau)d\tau d\bar\tau/\Im(\tau)$$

where D is a fundamental domain for the action of $SL_2(\mathbb{Z})$ on the upper half plane and f is a meromorphic modular form of weight $-s/2$ with poles only at cusps. The function $\Im(\tau)\Theta_L$ transforms like a modular form of weight $s/2$ and the differential $d\tau d\bar\tau/\Im(\tau)^2$ is invariant under $SL_2(\mathbb{Z})$ so the integrand is invariant under $SL_2(\mathbb{Z})$ and the integral does not depend on the choice of fundamental domain D. The integral needs to be interpreted rather carefully as it is wildly divergent because of terms of the form $e^{2\pi i n\tau}$ with $n < 0$. Harvey and Moore get round this by first integrating with respect to $\Re(\tau)$ and only then integrating with respect to $\Im(\tau)$ (although this still leaves a problem with integrating $1/\Im(\tau)$ from 1 to infinity, which they deal with by first subtracting a suitable function from the integrand).

The integral above is similar to the integral used by Niwa [Ni] in his work on the Shimura correspondence. This can be used to explain the formal similarities observed in [B95] between the Shimura correspondence and the infinite products above; see section 11 below. In particular Harvey and Moore's formula can be thought of as a sort of version of the Howe correspondence for the dual reductive pair $O_{s+2,2}(\mathbb{R})$ and $SL_2(\mathbb{R})$ for functions with singularities.

Harvey and Moore formally evaluate this integral and find that it is essentially given by the logarithm of the absolute value of the function Φ

of the theorem above, plus a few elementary factors which account for the fact that the integral is invariant under $Aut(II_{r,s})^+$ while Φ is not quite invariant.

Jorgenson and Todorov found a quite different way of constructing similar automorphic functions on moduli spaces as analytic discriminants [JT]. The relation to the work above is that the moduli spaces of Enriques surfaces and polarized K3 surfaces are (roughly) quotients of the hermitian symmetric spaces of lattices of signature $(n, 2)$ for $n = 10$ or 18. It seems likely that some of the functions constructed by Jorgensen and Todorov can also be constructed using some variation of Harvey and Moore's method.

Automorphic forms which are infinite products have recently turned up in several papers by physicists on string theory, but I do not understand this well enough to report on it. See [HM] and [D] for example.

10 Some superalgebras of rank 10

Harvey and Moore's formula for Φ has the big advantage that it generalizes easily to arbitrary lattices L of any dimension, signature, and determinant; see [B] for details. We will give a few examples in this section.

One extra feature that appears in the higher level case is that an automorphic form can have several apparently quite different infinite product expansions, converging in different regions. A classical example of this is the different product expansions for the theta function of a one dimensional lattice. If we put

$$\theta(\tau) = \sum_{n\in\mathbb{Z}}(-1)^n q^{n^2/2}$$

then it is a modular form for the group $\Gamma(2)$, and has the following infinite product expansions at cusps of $\Gamma(2)$.

$$\begin{aligned}\theta(\tau) &= 1 - 2q^{1/2} + 2q^{4/2} - 2q^{9/2} + \cdots \\ &= (1-q^{1/2})^2(1-q)(1-q^{3/2})^2(1-q^2)\cdots \\ (\tau/i)^{-1/2}\theta(-1/\tau) &= 2q^{1^2/8} + 2q^{3^2/8} + 2q^{5^2/8} + \cdots \\ &= 2q^{1/8}(1-q)^{-1}(1-q^2)(1-q^3)^{-1}(1-q^4)\cdots\end{aligned}$$

So two apparently quite different infinite products are really the same function expanded around different cusps.

We will take L to be the lattice $II_{9,1} \oplus II_{1,1}(2)$, where $II_{1,1}(2)$ is the lattice $II_{1,1}$ with all norms multiplied by 2. We define a vector valued modular form F with components f_{ij} for $i, j \in \mathbb{Z}/2\mathbb{Z}$ by

$$f_{00}(\tau) = 8\eta(2\tau)^8/\eta(\tau)^{16} = 8 + 128q + 1152q^2 + \cdots$$
$$f_{10}(\tau) = f_{01}(\tau) = -8\eta(2\tau)^8/\eta(\tau)^{16} = -8 - 128q - 1152q^2 - \cdots$$
$$f_{11}(\tau) = 8\eta(2\tau)^8/\eta(\tau)^{16} + \eta(\tau/2)^8/\eta(\tau)^{16} = q^{-1/2} + 36q^{1/2} + 402q^{3/2} + \cdots$$

Then there is an automorphic form such that the log of its absolute value is (more or less) given by

$$\int_D \Theta_L(\tau, \bar\tau, v)F(\tau)d\tau d\bar\tau/\Im(\tau)$$

(where the Siegel theta function is now a certain vector valued function taking values in the group ring of L'/L).

By theorem 13.3 of [B] we can find an infinite product expansion of this automorphic form for every primitive norm 0 vector of L (or equivalently to every one dimensional cusp). Define $c(n)$ by

$$\sum_n c(n)q^n = f_{00}(\tau) + f_{11}(\tau) = q^{-1/2} + 8 + 36q^{1/2} + O(q).$$

Then the infinite product at one cusp (for K the even sublattice of $I_{9,1}$) is

$$e^{2\pi i(\rho,v)} \prod_{\substack{\lambda\in K' \\ (\lambda,W)>0}} (1 - e^{2\pi i(v,\lambda)})^{\pm c(-\lambda^2/2)}$$
$$= \sum_{w\in G} \det(w)e^{2\pi i(w(\rho),v)} \prod_n (1 - e^{2\pi i n(w(\rho),v)})^{(-1)^n 8}$$

where the sign in the exponent is 1 if $\lambda \in K$ or if λ has odd norm, and 1 if λ has even norm but is not in K. The group G is the reflection group generated by reflections of the norm 1 vectors of K. The infinite product at the other cusp (with $K = II_{9,1}$) is

$$\prod_{\lambda\in K(\lambda,W)>0} (1 - e^{(v,\lambda)})^{c(-\lambda^2/2)}(1 + e^{(v,\lambda)})^{-c(-\lambda^2/2)} = 1 + \sum_\lambda a(\lambda)e^{2\pi i(v,\lambda)}$$

where $a(\lambda)$ is 1 if $\lambda = 0$, the coefficient of q^n of

$$\eta(\tau)^{16}/\eta(2\tau)^8$$

if λ is n times a primitive norm 0 vector in the closure of the positive cone C, and 0 otherwise.

In both cases we can identify the infinite product as a sum using the fact that it is an automorphic form of singular weight, so that all its Fourier coefficients corresponding to values of λ of nonzero norm vanish, and we get the formulas above. Both of these formulas are denominator formulas for generalized Kac-Moody superalgebras of rank 10 [R], and both superalgebras can be constructed as spaces of states of a superstring moving on a 10 dimensional torus.

The automorphic form can also be considered as a function on the period space of marked Enriques surfaces as in [B96], which can be used to show that the moduli space of Enriques surfaces is quasiaffine.

11 The Shimura correspondence

The Shimura correspondence is a map from modular forms of weight $k+1/2$ to modular forms of weight $2k$. Shimura's original definition in [S] was rather roundabout and involved taking an eigenform of weight $k+1/2$ under the Hecke operators, taking the corresponding Euler product, changing it in a mysterious way to a new Euler product, and then using Weil's theorem to reconstruct a modular form from this new Euler product. Niwa [Ni] and Kohnen [Ko] reformulated Shimura's result and found a more straightforward way of constructing Shimura's map. Combining their results in the level 1 case (and making an easy extension to the case $k=0$) we get the following theorem as a special case.

Theorem. *Suppose that $f(\tau)=\sum_n c(n)q^n$ is a modular form of weight $k+1/2>0$ for $\Gamma_0(4)$ (with k even) such that the Fourier coefficients $c(n)$ vanish unless $n\equiv 0,1 \bmod 4$. Then the function $\Phi(\tau)$ defined by*

$$\Phi(\tau)=-c(0)B_k/2k+\sum_{n\neq 0}q^n\sum_{0<d|n}d^{k-1}c(n^2/d^2)$$

is a modular form of weight $2k$ if $k>0$. If $k=0$ and all the coefficients $c(n)$ are integers then for some rational h the functions $q^h\exp(\Phi(\tau))$ is a modular form of weight $c(0)$ for some character of $SL_2(\mathbb{Z})$ (at least if we first remove the infinite constant term from the expression for Φ).

Example. If $k=0$ then we take $f(\tau)=\theta(\tau)=\sum_n q^{n^2}$. Then

$$\Phi(\tau)=\log\left(\prod_{n>0}(1-q^n)^2\right)$$

so that $q^h\exp(\Phi(\tau))$ is $\eta(\tau)^2$ and $h=1/12$.

Example. Put $f_{13/2} = \theta F(\theta^4 - 16F)(\theta^4 - 2F)$ of weight $6 + 1/2$ where $F(\tau) = \sum_{n>0} \sigma_1(2n+1)q^{2n+1}$. Then $\Phi(\tau)$ must be a form of weight 12, so must be (a multiple of) $\Delta(\tau)$. We find

$$\begin{aligned} f_{13/2}(\tau) &= q - 56q^4 + 120q^5 - 240q^8 + 9q^9 + 1440q^{12} - 1320q^{13} \\ &\quad - 704q^{16} + O(q^{17}) \\ \Delta(\tau) &= \sum_n \tau(n)q^n = q - 24q^2 + 252q^3 - 1472q^4 + O(q^5) \end{aligned}$$

and can check explicitly in the example above that

$$\tau(n) = \sum_{d|n} d^5 c(n^2/d^2).$$

Niwa showed that the function Φ could be obtained by considering the integral of $\bar{\Theta}(\tau)f(\tau)$, where Θ is the Siegel theta function of a 3 dimensional lattice. This is very similar to Harvey and Moore's formula in section 9, and in fact both formulas turn out to be special cases of a sort of Howe correspondence for modular forms, possibly with poles at cusps [B]. In particular the Shimura correspondence stated above still works perfectly well if the form f is allowed to have poles at cusps, except that the form Φ will now have singularities at imaginary quadratic irrationals coming from the poles of f. The singularities will be poles of order k if $k > 0$ and logarithmic singularities (corresponding to the poles and zeros of $q^h \exp(\Phi)$) if $k = 0$.

Example. Put

$$\begin{aligned} f(\tau) &= f_{13/2}(\tau)E_8(4\tau)/\Delta(4\tau) + 6720\sum_n H(2,n)q^n \\ &= q^{-3} + 64q - 32384q^4 + 131535q^5 - 4257024q^8 + 11535936q^9 + O(q^{12}) \\ &= \sum_n c(n)q^n \end{aligned}$$

(where $H(2,n) = L(-1,\chi_n)$ is Cohen's function [Co]) so that F has weight $5/2$. We see that $\Psi(\tau)$ should is a modular form of weight $2(5/2 - 1/2) = 4$, and we can work out the singularities and find that they are poles of order 2 at the conjugates of a cube root of 1. We find that $\Psi(\tau)$ is

$$\begin{aligned} 64\Delta(\tau)/E_4(\tau)^2 &= 64(q - 504q^2 + 180252q^3 - 56364992q^4 + O(q^5)) \\ &= \sum_n q^n \sum_{d|n} dc(n^2/d^2) \end{aligned}$$

which has poles of order 2 at cube roots of 1 and their conjugates because E_4 has zeros of order 1 at these points.

Example. Now we look at an example where $k = 0$ and f has poles at the cusp. We take

$$\begin{aligned} f(\tau) &= -2f_{13/2}(\tau)E_6(4\tau)/\Delta(4\tau) - 108\theta(\tau) \\ &= -2q^{-3} + 4 + 504q - 53496q^4 + 171990q^5 \\ &\quad - 3414528q^8 + 8192504q^9 + O(q^{12}) \\ &= \sum_n c(n)q^n. \end{aligned}$$

This time $q^h \exp(\Psi)$ should again have weight equal to the constant term 4, and also turns out to have poles of order 2 at conjugates of a cube root of 1. These poles come from the term $-2q^{-3}$ as follows: the coefficient -2 is the order of the zero, and the exponent 3 is the discriminant of the quadratic equations whose roots are conjugates of a cube root of 1. We find that

$$\begin{aligned} q^h \exp(\Psi) &= \Delta(\tau)/E_4(\tau)^2 \\ &= q \prod_{n>0} (1 - q^n)^{c(n^2)} \end{aligned}$$

is the same as the function in the previous example up to a factor of 64. So both the coefficients of this function and the exponents in its infinite product expansion can be given in terms of modular forms of half integral weight with poles at the cusps.

As we already know an infinite product for Δ we can use the example above to find an explicit infinite product for the Eisenstein series $E_4(\tau)$ and hence also for the elliptic modular function $j(\tau) = E_4(\tau)^3/\Delta(\tau)$. More generally we can find an infinite product expansion for any level 1 modular function with integral coefficients all of whose zeros and poles are at imaginary quadratic irrationals or cusps.

12 Finiteness theorems

One can ask whether the examples discussed above are isolated and exceptional objects or whether they are part of an infinite family. It is possible to generalize most of the constructions above to produce a few hundred similar examples of things similar to (say) the fake monster Lie algebra. However there are several theorems suggesting that the total number of examples like this may be finite.

The nice behavior of the monster Lie algebra depends on the fact that the reflection group of $II_{25,1}$ is very nice, and in particular on the fact that it has a (Weyl) vector ρ which has bounded inner product with all simple roots. If such a vector exists in some Lorentzian lattice L with negative norm then there are only a finite number of simple roots of L and the reflection group of L has finite index in the full automorphism group. V. Nikulin has shown [N] that there are essentially only a finite number of such lattices, up to multiplication by constants, and Esselmann has shown that the only one of dimension greater than 20 is the 22 dimensional lattice of determinant 4 consisting of the even vectors of $I_{21,1}$. Nikulin recently extended his theorem to cover the case when the vector ρ has zero norm. The largest lattice in this case is presumably the lattice $II_{25,1}$ though this has not been proved. So Nikulin's work suggests that there may be only a finite number of interesting Lie algebras similar to the monster Lie algebra. However there are some examples of generalized Kac-Moody algebras with no real roots which still have known simple roots and root multiplicities and Weyl vectors equal to 0, so it is still conceivable (but unlikely) that there are an infinite number of these. Nikulin and Gritsenko have given some examples of generalized Kac-Moody algebras related to some hyperbolic reflection groups in [GN] and have suggested that maybe most crystallographic reflection groups in Lorentzian lattices L which have finite index in the full automorphism group are associated with some nice generalized Kac-Moody algebra or superalgebra whose denominator formula is an automorphic form. In particular they show that the Siegel modular form Δ_5 (one of the standard generators of the ring of Siegel modular forms of genus 2) can be written as such an infinite product.

Many of the Lie algebras similar to the fake monster Lie algebra (and all the known ones of rank at least 3) are closely related to certain products of (positive and negative) powers of η functions with multiplicative coefficients. For example the fake monster Lie algebra itself is related to the function $\Delta(\tau) = \eta(\tau)^{24}$, whose coefficients $\tau(n)$ were proved to be multiplicative by Mordell and Hecke. Y. Martin [M] has recently found all such products of eta functions and in particular has proved that there are only a finite number of them. This again hints that there are only finitely many analogues of the fake monster Lie algebra.

Some of the generalized Kac-Moody algebras of rank 2, such as the monster Lie algebra itself, are not related to multiplicative products of eta functions. However most of the known ones that are not are instead closely related to certain Hauptmoduls of genus 0 congruence subgroups of $SL_2(\mathbb{R})$; for example, the monster Lie algebra is related to the Hauptmodul $j(\tau)$ of the subgroup $SL_2(\mathbb{Z})$. It is easy to find an infinite number of non congruence genus 0 subgroups, but J. G. Thompson showed that there are only a finite number of conjugacy classes of congruence subgroups of any given genus.

The theorem in section 9 produces an infinite supply of examples of automorphic forms with infinite product expansions. Unfortunately most of these cannot possibly be the denominator formulas of generalized Kac-Moody algebras. The point is that most of the infinite products involve vectors of norm at least 4 in the lattice, so such vectors would have to be roots of any generalized Kac-Moody algebra. However if v is a positive root of a generalized Kac-Moody algebra, then $(v,v)|2(v,w)$ for any other root w. This means that if the root lattice is unimodular then positive norm vectors cannot have norms greater than 2. In fact the only infinite products given by the theorem in section 9 which can be the denominator formulas of generalized Kac-Moody algebras are the ones associated to the fake monster Lie algebra and the monster Lie algebra that we have already seen above. There are of course many automorphic forms of higher level which have nice infinite product expansions, and it seems conceivable that there may be infinite families of these with no positive norm vectors appearing in the infinite product. These could then be interpreted as the denominator formulas for generalized Kac-Moody superalgebras with known differences of the root multiplicities and simple root multiplicities.

Bibliography

[B86] R. E. Borcherds, Vertex algebras, Kac-Moody algebras, and the monster. Proc. Natl. Acad. Sci. USA. Vol. 83 (1986) 3068-3071.

[B90] R. E. Borcherds, The monster Lie algebra, Adv. Math. Vol. 83, No. 1, Sept. 1990.

[B92] R. E. Borcherds, Monstrous moonshine and monstrous Lie superalgebras, Invent. Math. 109, 405-444 (1992).

[B94] R. E. Borcherds, Simple groups and string theory, First European congress of mathematics, Paris July 1992, Ed. A. Joseph and others,vol. 1 p. 411-421, Birkhauser 1994.

[B95] R. E. Borcherds, Automorphic forms on $O_{s+2,2}(R)$ and infinite products. Invent. Math. 120, p. 161-213 (1995)

[B96] R. E. Borcherds, The moduli space of Enriques surfaces and the fake monster Lie superalgebra. Topology vol. 35 no. 3, 699-710, 1996.

[B] R. E. Borcherds, Automorphic forms with singularities on Grassmannians, preprint alg-geom/9609022, To appear in Invent. Math. 1998.

[Co] H. Cohen, Sums involving the values at negative integers of L-functions of quadratic characters. Math. Ann. 217, 271-285 (1975).

[C] J. H. Conway, The automorphism group of the 26 dimensional even Lorentzian lattice. J. Algebra 80 (1983) 159-163. This paper is reprinted as chapter 27 of [CS].

[CS] J. H. Conway, N. J. A. Sloane. Sphere packings, lattices and groups. Springer-Verlag New York 1988, Grundlehren der mathematischen Wissenschaften 290.

[D] R. Dijkgraaf, E. Verlinde and H. Verlinde, Counting dyons in $N = 4$ string theory. Nuclear Phys. B 484 (1997), no. 3, 543-561, hep-th/9607026, and R. Dijkgraaf, G. Moore, E. Verlinde and H. Verlinde, Elliptic genera of symmetric products and second quantized strings. Comm. Math. Phys. 185 (1997), no. 1, 197-209, hep-th/9608096.

[E-Z] M. Eichler, D. Zagier, "The theory of Jacobi forms", Progress in mathematics vol. 55, Birkhäuser, Boston, Basel, Stuttgart 1985.

[F] I. Frenkel, Representations of Kac-Moody algebras and dual resonance models. Applications of group theory in physics and mathematical physics, Chicago 1982, p. 325-353, Lectures in applied math. 21, A.M.S.

[FF] I. Frenkel, A. Feingold, A hyperbolic Kac-Moody algebra and the theory of Siegel modular forms of genus 2, Math. Ann. 263 (1983) no. 1, 87-144.

[FLM] I. B. Frenkel, J. Lepowsky, A. Meurman, Vertex operator algebras and the monster, Academic press 1988.

[FGZ] I. B. Frenkel, H. Garland, G. Zuckerman, Semi-infinite cohomology and string theory, Proc. Natl. Acad. Sci. U.S.A. 83 (1986), 8442-8446.

[G] R. W. Gebert, Introduction to vertex algebras, Borcherds algebras and the monster Lie algebra, Internat. J. Modern Physics A8 (1993) no 31, 5441-5503.

[Gr] V. A. Gritsenko, Jacobi functions of n variables. J. Soviet Math. 53 (1991), 243-252. (The original Russian is in Zap. Nauch. Seminars of LOMI v. 168 (1988), p. 32-45.)

[GN] V. A. Gritsenko, V. V. Nikulin, Siegel automorphic form corrections of some Lorentzian Kac-Moody algebras, Amer. J. Math. 119 (1997), no. 1, 181-224.

[HM] J. Harvey, G. Moore, Algebras, BPS states, and strings. Nuclear Phys. B 463 (1996), no. 2-3, 315-368, hep-th/9510182.

[J] E. Jurisich, Generalized Kac-Moody Lie algebras, free Lie algebras and the structure of the Monster Lie algebra. To appear in Journal of pure and applied algebra.

[JT] J. Jorgenson, A. Todorov, Enriques surfaces and analytic discriminants, preprint 1995. An analytic discriminant for polarized K3 surfaces, 1994 preprint.

[K] V. G. Kac, "Infinite dimensional Lie algebras", third edition, Cambridge University Press, 1990.

[K95] V. G. Kac, Vertex algebras for beginners. University Lecture Series, 10. American Mathematical Society, Providence, RI, 1997. viii+141 pp. ISBN: 0-8218-0643-2.

[KW] V. Kac, M. Wakimoto, Integrable highest weight modules over affine superalgebras and number theory, in "Lie theory and geometry in honor of Bertram Kostant" p. 415-456, Editors J.-L. Brylinski, R. Brylinski, V. Guillemin, V, Kac, Progress in mathematics vol. 123, Birkhauser 1994.

[Ko] W. Kohnen, Modular forms of half integral weight on $\Gamma_0(4)$, Math. Ann. 248, 249-266 (1980).

[LZ] B. Lian, G. Zuckerman, New perspectives on the BRST-algebraic structure in string theory, hepth/921107, Communications in Mathematical Physics 154, (1993) 613-64, and Moonshine cohomology, q-alg/ 950101 Finite groups and Vertex Operator Algebras, RIMS publication (1995) 87-11.

[M] Y. Martin, Multiplicative η-quotients. Trans. Amer. Math. Soc. 348 (1996), no. 12, 4825-4856. On Hecke operators and products of the Dedekind η-function. C. R. Acad. Sci. Paris Sér. I Math. 322 (1996), no. 4.

[Nie] P. Niemann, On some generalized Kac-Moody algebras with known root multiplicities, Cambridge PhD Thesis, 1994.

[N] V. V. Nikulin, Discrete reflection groups in Lobachevsky spaces and algebraic surfaces, Proceedings of the I.C.M., Berkeley, 1986, 654-671.

[Ni] S. Niwa, Modular forms of half integral weight and the integral of certain theta functions, Nagoya Math J. 56 (1975), 147-163.

[R] U. Ray, A character formula for generalized Kac-Moody superalgebras. J. Algebra 177 (1995), no. 1, 154-163.

[S] G. Shimura, On modular forms of half integral weight, Ann. of Math. 97, 440-481 (1973).

[Y] J.-H. Yang, Kac-Moody algebras, the monstrous moonshine, Jacobi forms, and infinite products, p. 13-82, Proceedings of the 1995 symposium on number theory geometry and related topics, Editors J.-W. Son, J.-H. Yang, The Pyoungsan institute for the mathematical sciences, May 1996.

Harmonic analysis for affine Hecke algebras

G.J. Heckman
Department of Mathematics,
University of Nijmegen,
Toernooiveld, 6525 ED Nijmegen,
The Netherlands
E-mail: heckman@sci.kun.nl

E.M. Opdam
Department of Mathematics,
University of Leiden,
P.O. Box 9512, 2300 RA Leiden,
The Netherlands
E-mail: opdam@wi.leidenuniv.nl

Recently the classification of unipotent representations of a simple p-adic group was obtained by Lusztig based on his ideas of character sheaves. An alternative strategy had been proposed by Reeder based on a comparison of formal degrees. We show how the formal degrees can be computed from a residue calculation and thereby turn the approach of Reeder into an efficient route.

1 The affine Hecke algebra

Suppose V is a real vector space of dimension n equipped with a positive definite scalar product $(\cdot,\cdot)$. Let $R \subset V$ be a reduced irreducible root system, and write $R^\vee = \{\alpha^\vee; \alpha \in R\}$ for the dual root system. Here

$$\alpha^\vee = \frac{2\alpha}{(\alpha,\alpha)}$$

is the coroot of a root $\alpha \in R$. A root $\alpha \in R$ will be considered as a linear function on V by $\alpha(\xi) = (\xi, \alpha)$ for $\xi \in V$. Let δ denote the constant function 1 on V. The set R' of affine linear functions on V defined by

$$R' = \{\alpha + m\delta; \alpha \in R, m \in \mathbf{Z}\} \tag{1.1}$$

is called the *affine root system* associated with R.
For each affine root $a \in R'$ let H_a denote the affine hyperplane in V on which a vanishes, and let s_a denote the orthogonal reflection in H_a. Explicitly

$$s_{\alpha+m\delta}(\xi) = \xi - ((\xi,\alpha) + m)\alpha^\vee$$

for $a = \alpha + m\delta \in R'$ and $\xi \in V$. The *affine Weyl group* W' associated with R is the group of euclidean motions of V generated by these reflections. It contains the finite Weyl group W_0 of R (and of $R^\vee$) as the subgroup generated by the reflections s_α for $\alpha \in R$. For each $\alpha \in R$ the product $s_\alpha s_{\alpha+\delta}$ maps $\xi \in V$ to $\xi + \alpha^\vee$, and therefore

$$t(\alpha^\vee) = s_\alpha s_{\alpha+\delta}$$

is translation of V over $\alpha^\vee$. It follows that W' contains the subgroup of translations over elements of the coroot lattice $Q^\vee = \mathbf{Z}R^\vee$ of R. In fact we have a semidirect product decomposition

$$W' = W_0 \ltimes t(Q^\vee). \tag{1.2}$$

For our purpose it will be more convenient to work with the *extended affine Weyl group*

$$W = W_0 \ltimes t(P^\vee) \tag{1.3}$$

with $P^\vee = \{\lambda \in V; (\lambda,\alpha) \in \mathbb{Z}\ \forall \alpha \in R\}$ the coweight lattice of R. It is easy to check that W contains W' as normal subgroup. The quotient W/W' is isomorphic to the finite abelian group $P^\vee/Q^\vee$.

Fix a Weyl chamber V_+ for R. Let R_+ be the set of positive roots relative to V_+, and let $\alpha_1, \ldots, \alpha_n \in R_+$ be the set of simple roots. Let $\theta \in R_+$ be the highest root. The affine roots

$$a_0 = -\theta + \delta, \; a_i = \alpha_i \quad (1 \leq i \leq n) \tag{1.4}$$

form a set of simple roots for R', and the affine Weyl group W' is generated by the reflections $s_i = s_{a_i}$ $(0 \leq i \leq n)$. The alcove

$$C = \{\xi \in V; a_i(\xi) > 0 \; (0 \leq i \leq n)\} \tag{1.5}$$

is the unique connected component of $V \backslash \cup H_a$ (union over $a \in R'$) which contains the origin in its closure and is contained in V_+. The affine Weyl group W' permutes the connected components of $V \backslash \cup H_a$ in a simply transitive way, and therefore yields a tessellation of V by congruent simplices. Likewise the extended affine Weyl group W permutes the connected components of $V \backslash \cup H_a$ in a transitive way, and therefore we get a semidirect product decomposition

$$W = \Omega \ltimes W' \tag{1.6}$$

with

$$\Omega = \{w \in W; w(C) = C\} \cong P^\vee / Q^\vee. \tag{1.7}$$

For $w \in W$ the length $l(w) \in \mathbb{N}$ is defined as the number of hyperplanes H_a $(a \in R'_+)$ separating the two alcoves C and wC. Here $R'_+ = \{a \in R'; a(\xi) > 0 \; \forall \xi \in C\}$ is the set of positive affine roots. It is easy to check that

$$R'_+ = R_+ \cup \{R + (\mathbb{N} + 1)\delta\}. \tag{1.8}$$

We write

$$\rho = \frac{1}{2} \sum_{\alpha \in R_+} \alpha \tag{1.9}$$

and denote

$$P_+^\vee = \{\lambda \in P^\vee; (\lambda, \alpha) \in \mathbb{N} \quad \forall \alpha \in R_+\} \tag{1.10}$$

for the cone of dominant coweights. Then it is easy to show that

$$l(t(\lambda)) = 2(\lambda, \rho) \qquad \text{for } \lambda \in P_+^\vee. \tag{1.11}$$

Let $q \in \mathbb{C}^\times$ be a nonzero complex parameter. The *extended affine Hecke algebra* $\mathbf{H} = \mathbf{H}(W, q)$ is the associative algebra over $\mathbb{C}$ with a vector space basis T_w indexed by $w \in W$, and multiplication given by

$$(T_i + 1)(T_i - q) = 0 \qquad \text{for } 0 \le i \le n, \tag{1.12}$$

$$T_v T_w = T_{vw} \qquad \text{if } l(v) + l(w) = l(vw). \tag{1.13}$$

Here we abbreviate T_{s_i} by T_i. The subspace of $\mathbf{H}$ with basis T_w $(w \in W')$ is a subalgebra called the *affine Hecke algebra* and denoted $\mathbf{H}' = \mathbf{H}(W', q)$. The subspace of $\mathbf{H}$ with basis T_ω $(\omega \in \Omega)$ is a subalgebra naturally isomorphic to $\mathbb{C}[\Omega]$. The semidirect product (1.6) gives rise to the isomorphism

$$\mathbf{H} \cong \mathbb{C}[\Omega] \otimes \mathbf{H}' \tag{1.14}$$

with $T_v T_\omega = T_\omega T_w$ if $\omega w = v\omega$ $(v, w \in W', \omega \in \Omega)$. The subspace of $\mathbf{H}$ with basis T_w $(w \in W_0)$ is a subalgebra denoted $\mathbf{H}_0 = \mathbf{H}(W_0, q)$. This is the Hecke algebra of the finite Weyl group W_0.

From (1.11) and (1.13) it is clear that

$$T_{t(\lambda)} T_{t(\mu)} = T_{t(\lambda+\mu)} \qquad \text{for } \lambda, \mu \in P_+^\vee. \tag{1.15}$$

If now λ is any element of $P^\vee$ then we can write $\lambda = \mu - \nu$ with $\mu, \nu \in P_+^\vee$, and we define

$$\theta_\lambda = q^{-(\lambda,\rho)} T_{t(\mu)} T_{t(\nu)}^{-1}. \tag{1.16}$$

Using (1.15) it is easy to see that this definition is unambiguous, and in addition

$$\theta_\lambda \theta_\mu = \theta_{\lambda+\mu} \qquad \text{for all } \lambda, \mu \in P^\vee. \tag{1.17}$$

Hence the subspace of $\mathbf{H}$ with basis θ_λ $(\lambda \in P^\vee)$ is a subalgebra naturally isomorphic to $\mathbb{C}[P^\vee]$. Next one can show that multiplication in $\mathbf{H}$ defines a vector space isomorphism

$$\mathbf{H} \cong \mathbf{H}_0 \otimes \mathbb{C}[P^\vee]. \tag{1.18}$$

Moreover the algebra structure on $\mathbf{H}$ can be recovered from the algebra structures on $\mathbf{H}_0$ and $\mathbb{C}[P^\vee]$ together with the push relation

$$\theta_\lambda T_i - T_i \theta_{s_i(\lambda)} = (q-1)\frac{\theta_\lambda - \theta_{s_i(\lambda)}}{1 - \theta_{-\alpha_i^\vee}} \tag{1.19}$$

for $\lambda \in P^\vee$ and $1 \leq i \leq n$. Note that the division on the right hand side is possible inside $\mathbb{C}[P^\vee]$. This formula is due to Lusztig [Lu3, Mac1] but on the level of braid groups it was obtained before by van der Lek and Looijenga [Le1,2]. Using (1.19) it is easy to see that the center $\mathbf{Z}$ of $\mathbf{H}$ is equal to

$$\mathbf{Z} = \mathbb{C}[P^\vee]^{W_0} \tag{1.20}$$

with $\mathbb{C}[P^\vee]^{W_0}$ the subalgebra of $\mathbb{C}[P^\vee]$ of invariants for W_0. Here W_0 acts on $\mathbb{C}[P^\vee]$ by $w(\theta_\lambda) = \theta_{w(\lambda)}$ for $w \in W_0$ and $\lambda \in P^\vee$.

Let $P_{W_0}(q) = \sum_{w \in W_0} q^{l(w)}$ be the Poincaré polynomial of W_0. Assume from now on that $P_{W_0}(q) \neq 0$, and let

$$T_0^+ = P_{W_0}(q)^{-1}\left(\sum_{w \in W_0} T_w\right) \in \mathbf{H}_0 \tag{1.21}$$

be the central idempotent corresponding to the trivial representation $triv$: $\mathbf{H}_0 \to \mathbb{C}$, $triv(T_w) = q^{l(w)}$. The Satake isomorphism expresses that multiplication by T_0^+ defines an isomorphism of commutative algebras

$$\mathbf{Z} = \mathbb{C}[P^\vee]^{W_0} \xrightarrow{\cong} T_0^+ \mathbf{H} T_0^+,\ \varphi \mapsto T_0^+\varphi. \tag{1.22}$$

The spherical subalgebra $T_0^+\mathbf{H}T_0^+$ of $\mathbf{H}$ has the vector space basis $T_0^+\theta_\lambda T_0^+$ with $\lambda \in P_+^\vee$. An explicit inversion for the Satake isomorphism was obtained by Macdonald [Mac2] in the form

$$T_0^+\theta_\lambda T_0^+ = P_{W_0}(q^{-1})^{-1} T_0^+ \left(\sum_{w \in W_0} w(c(\cdot, q)\theta_\lambda)\right) \tag{1.23}$$

for $\lambda \in P_+^\vee$, and with the c-function given by

$$c(\cdot, q) = \prod_{\alpha \in R_+} \frac{1 - q^{-1}\theta_{-\alpha^\vee}}{1 - \theta_{-\alpha^\vee}}. \tag{1.24}$$

Note that although $c(\cdot, q)$ is no longer an element of $\mathbb{C}[P^\vee]$ the averaging over W_0 in the right hand side of (1.23) makes the outcome lie in $\mathbb{C}[P^\vee]^{W_0}$.

The irreducible representations of $\mathbb{C}[P^\vee]$ are parametrized by the complex torus $T = \mathrm{Hom}(P^\vee, \mathbb{C}^\times)$ with character lattice $P^\vee$. For $s \in T$ the induced representation

$$\mathrm{Ind}_{\mathbb{C}[P^\vee]}^{\mathbf{H}}(s) \tag{1.25}$$

is called the *principal series* representation of $\mathbf{H}$ with spectral parameter s. As a module for $\mathbf{H}_0$ it is just the left regular action of $\mathbf{H}_0$ by (1.18). For regular $s \in T$ it is as a module for $\mathbb{C}[P^\vee]$ the direct sum of the one dimensional modules ws ($w \in W_0$). As shown by Kato [Ka] the principal series (1.25) is irreducible if and only if

$$\text{numerator}(c(s,q)c(s^{-1},q)) \neq 0. \tag{1.26}$$

Each irreducible representation of $\mathbf{H}$ is equivalent to a quotient of some principal series representation.

2 The spherical Plancherel formula

In this section we take $q > 0$ a positive real parameter. The extended affine Hecke algebra $\mathbf{H} = \mathbf{H}(W, q)$ becomes an involutive algebra with respect to

$$(\sum_w c_w T_w)^* = \sum_w \bar{c}_{w^{-1}} T_w. \tag{2.1}$$

The adjoint is an antilinear antiinvolution of $\mathbf{H}$. The canonical trace τ : $\mathbf{H} \to \mathbb{C}$ is defined by

$$\tau(\sum_w c_w T_w) = c_e, \tag{2.2}$$

and it gives rise to the canonical hermitian inner product

$$(\varphi, \psi) = \tau(\varphi\psi^*) = \sum_w c_w \bar{d}_w q^{l(w)} \tag{2.3}$$

with $\varphi = \sum c_w T_w$, $\psi = \sum d_w T_w \in \mathbf{H}$. Denote by $\mathbf{H}^\wedge$ the set of equivalence classes of irreducible unitary representations of $\mathbf{H}$. The abstract Plancherel theorem [Di] for $\mathbf{H}$ gives the existence of a unique nonnegative measure $\nu_P(\cdot) = \nu_P(\cdot, q)$ on $\mathbf{H}^\wedge$ such that

$$\tau(\varphi) = \int_{\pi \in \mathbf{H}^\wedge} Tr(\pi(\varphi)) d\nu_P(\pi, q) \quad \forall \varphi \in \mathbf{H}. \tag{2.4}$$

The measure ν_P on $\mathbf{H}^\wedge$ is called the canonical *Plancherel measure* for $\mathbf{H}$.

Let $\mathbf{C} = \mathbf{C}(W, q)$ be the formal completion of $\mathbf{H}$ with respect to the basis T_w ($w \in W$). Hence $\varphi \in \mathbf{C}$ means that $\varphi = \sum c_w T_w$ is a formal infinite sum with complex coefficients. Clearly for $\varphi \in \mathbf{C}$ and $\psi \in \mathbf{H}$ the product $\varphi\psi \in \mathbf{C}$ and the hermitian pairing $(\varphi, \psi) \in \mathbb{C}$ are well defined. The subspace $T_0^+ \mathbf{C} T_0^+$ is the space of all spherical functions in $\mathbf{C}$. Now suppose in relation (2.4) that $\varphi \in T_0^+ \mathbf{H} T_0^+$. Clearly $Tr(\pi(\varphi)) = 0$ in the right hand side of (2.4) unless $\pi \in \mathbf{H}^\wedge_{sph}$ with

$$\mathbf{H}^\wedge_{sph} = \{\pi \in \mathbf{H}^\wedge; [\pi|_{\mathbf{H}_0} : triv] = 1\}. \tag{2.5}$$

For $\pi \in \mathbf{H}^\wedge_{sph}$ the *elementary spherical function* $\varphi_\pi \in T_0^+ \mathbf{C} T_0^+$ is uniquely characterized by

$$(\varphi_\pi, \varphi^*) = Tr(\pi(\varphi)) \qquad \forall \varphi \in T_0^+ \mathbf{H} T_0^+. \tag{2.6}$$

It is easy to check that

$$\varphi \varphi_\pi = (\varphi_\pi, \varphi^*) \varphi_\pi \quad \forall \varphi \in T_0^+ \mathbf{H} T_0^+. \tag{2.7}$$

The scalar (2.6) is called the *spherical Fourier transform* of φ (evaluated at π), and is also denotes $\widehat{\varphi}(\pi)$. The *spherical Plancherel formula* states that for all $\varphi, \psi \in T_0^+ \mathbf{H} T_0^+$

$$(\varphi, \psi) = \int_{\pi \in \mathbf{H}^\wedge_{sph}} \widehat{\varphi}(\pi)(\varphi_\pi, \psi) d\nu_P^+(\pi, q) \tag{2.8}$$

with ν_P^+ the restriction of the Plancherel measure ν_P to $\mathbf{H}^\wedge_{sph}$.

For $s \in T = \mathrm{Hom}(P^\vee, \mathbb{C}^\times)$ let π_s be the unique spherical subquotient of the principal series representation (1.25) with spectral parameter s. An immediate consequence of (1.23) is the explicit formula of Macdonald [Mac2, Theorem 4.1.2]

$$(\varphi_s, \theta_\lambda^*) = P_{W_0}(q^{-1})^{-1} \sum_{w \in W_0} c(ws, q) ws(\theta_\lambda) \tag{2.9}$$

for $\lambda \in P_+^\vee$, $s \in T$ regular for W_0, and φ_s short for φ_{π_s}. In case $q > 1$ the spherical Plancherel measure ν_P^+ has also been determined by Macdonald [Mac2, Theorem 5.1.2] in the form

$$d\nu_P^+(s, q) = \frac{q^{-N} ds}{|W_0| c(s, q) c(s^{-1}, q)} \tag{2.10}$$

with $N = \# R_+$ and ds the normalized Haar measure on the compact form T_c of T. The difference by a factor $P_{W_0}(q)$ between [Mac2, Theorem 5.1.2] and

(2.10) comes from a different normalization of Haar measure. Substituting $\psi = \theta_\lambda^*$ in (2.8) and using (2.9) and (2.10) yields (still for $q > 1$)

$$\begin{aligned}(\varphi, \theta_\lambda^*) &= \int_{T_c} \widehat{\varphi}(s)(\varphi_s, \theta_\lambda^*) d\nu_P^+(s, q) \\ &= P_{W_0}(q)^{-1} \int_{T_c} \widehat{\varphi}(s) s(\theta_\lambda) \frac{ds}{c(s^{-1}, q)}\end{aligned}$$

for $\lambda \in P_+^\vee$ and $\widehat{\varphi}(s) = \widehat{\varphi}(\pi_s)$. The integrand is meromorphic on T with simple poles along submanifolds of the form $\{t \in T; t(\theta_{\alpha^\vee}) = q\}$ for $\alpha \in R_+$. Therefore the contour of integration T_c can be shifted in the direction of the negative chamber

$$(\varphi, \theta_\lambda^*) = P_{W_0}(q)^{-1} \int_{s_0 T_c} \widehat{\varphi}(s) s(\theta_\lambda) \frac{ds}{c(s^{-1}, q)}, \tag{2.11}$$

or more precisely as long as $s_0 \in T_v = \{t \in T; t(\theta_\lambda) > 0 \ \forall \lambda \in P^\vee\}$ satisfies

$$s_0(\theta_{\alpha^\vee}) < q \qquad \text{for all } \alpha \in R_+. \tag{2.12}$$

Since both sides of (2.11) are analytic in q formula (2.11) therefore remains valid for all $q > 0$ as long as s_0 satisfies (2.12). In order to give (2.11) an L^2-interpretation for $0 < q < 1$ the contour $s_0 T_c$ should be shifted back at the cost of picking up residues. Grouping residues together according to orbits of W_0 enables one to compare the residue contributions in (2.11) with the ones obtained by a contour shift in the integral

$$\int_{s_0 T_c} \widehat{\varphi}(s)(\varphi_s, \theta_\lambda^*) d\nu_P^+(s, q). \tag{2.13}$$

The term $\widehat{\varphi}(s)(\varphi_s, \theta_\lambda^*)$ is holomorphic as function of s on all of T, and therefore the residues all come from the poles in the spherical Plancherel measure (2.10).

A connected algebraic subgroup of T is called a subtorus, and we will use the phrase "affine subtorus" (by analogy with the concept affine subspace of a vector space) for any subvariety S of T for which $s^{-1}S$ ($s \in S$) is a subtorus of T.

Definition 1 If for $S \subset T$ an affine subtorus we write

$$z(S) = \#\{\alpha \in R; s(\theta_{\alpha^\vee}) = 1 \quad \forall s \in S\} \tag{2.14}$$

$$p(S) = \#\{\alpha \in R; s(\theta_{\alpha^\vee}) = q \quad \forall s \in S\} \tag{2.15}$$

for the number of zeros and poles along S in the analytic continuation of the spherical Plancherel measure (2.10) then S is called residual if

$$p(S) \geq z(S) + codim(S). \tag{2.16}$$

Theorem 2.1 *For S a residual affine subtorus of T one always has the equality*

$$p(S) = z(S) + codim(S). \tag{2.17}$$

The proof of this theorem reduces immediately to the case that the affine subtorus S is just a point of T. Moreover one can classify the finite list of residual points of T for each root system case by case, and thereby verify (2.17) by inspection of the list. However the deeper reason for (2.17) is that it explains why under the contour shift in (2.13) measures are picked up rather than just distributions (as should in accordance with the abstract Plancherel theorem).

Theorem 2.2 *For $S \subset T$ a residual affine subtorus let*

$$S_c = \{s \in S;\ \textit{distance from } s \textit{ to } T_c \textit{ is minimal}\}. \tag{2.18}$$

Here the distance is taken with respect to a translation invariant metric on T, which is also invariant under W_0. Then S_c is a compact real form of S, and let μ_S denote the normalized invariant measure on S_c. For $0 < q < 1$ the spherical Plancherel measure $d\nu_P^+(s,q)$ is the measure on T given by

$$d\nu_P^+(s,q) = \sum_S c_S q^{-N} \frac{\Pi'|1 - s(\theta_{\alpha^\vee})|}{\Pi'|1 - q^{-1}s(\theta_{\alpha^\vee})|} d\mu_S(s) \tag{2.19}$$

with the sum over all residual affine subtori S of T. Here c_S is a nonnegative rational number with $c_{wS} = c_S$ for $w \in W_0$, and Π' denotes the product over all nonzero factors.

Remark 2.4. Note that $\mathbb{R}_{>0}$ acts on $T = T_v T_c$ by homotheties in T_v. The dependence of a residual affine subtorus S of T on q is simply a scale factor. The rationality of c_S therefore implies that c_S is independent of $q \in (0,1)$. All complications of the residue calculation are captured in the rational constants c_S. The actual computation of c_S can be difficult with complexity being exponential in $z(S)$. For $z(S) = 0$ it is easy and for $z(S) = 2$ it is manageable. By our method it is equally hard to decide whether $c_S > 0$

or $c_S = 0$, and therefore to conclude whether S_c really lies in the tempered spherical spectrum or not. However in the next section we will show using the work of Kazhdan and Lusztig [KL] that in fact

$$c_S > 0 \tag{2.20}$$

for each residual affine subtorus S of T.

Remark 2.5. For reasons of minimizing technicalities we have restricted ourselves to Hecke algebras $\mathbf{H}(W, q)$ with a single parameter q. However one can relax the quadratic relation (1.12) by requiring

$$(T_i + 1)(T_i - q^{k_i}) = 0 \tag{2.21}$$

instead. Here k_i are natural numbers satisfying $k_i = k_j$ if s_i and s_j are conjugated inside W. These multilabel Hecke algebras $\mathbf{H}(W, q^{\underline{k}})$ are important for the representation theory of reductive groups over a p-adic field. Virtually without change the results of this section go through for these Hecke algebras $\mathbf{H}(W, q^{\underline{k}})$. However in case of multilabel Hecke algebras it may happen in contrast with (2.20) that for $0 < q < 1$ and some $\underline{k} = (k_i)$

$$c_S = 0 \tag{2.22}$$

for some residual affine subtori S of T [HO, Proposition 4.16 or Table 4.18].

Remark 2.6. It is quite likely that the method discussed in this section can be extended (by working with arbitrary matrix coefficients rather than just spherical ones) to recover the full Plancherel formula in more explicit terms. An additional complication in comparison with the spherical case is that the residue contributions need no longer be single irreducible representations but finite collections (called packets) of these. All members of the same tempered packet should differ in Plancherel measure only by numbers independent of q.

Remark 2.7. The idea of picking up discrete spectrum from a residue calculation in the analytic continuation of the spectral measure of the continuous spectrum is familiar, and can be traced back to H. Weyl [We] in his treatment of the spectral problem for the hypergeometric function. It has been used by Selberg [Se] to find residual discrete spectrum in the meromorphic continuation of Eisenstein series. In several variables this method was used by Macdonald [Mac2, Section 5.2] and Matsumoto [Mat] in particular cases, and in full generality by Langlands [La] in the context of automorphic forms. A detailed exposition of this work of Langlands has been given by Moeglin and Waldspurger [MW].

3 The classification of Kazhdan and Lusztig

The classification of the irreducible representations of the unilabel extended affine Hecke algebras $\mathbf{H}(W,q)$ is quite well understood from the work of Kazhdan and Lusztig [KL]. The complex torus

$$T = \mathrm{Hom}(P^\vee, \mathbb{C}^\times) \tag{3.1}$$

parametrizing principal series representations (1.25) of $\mathbf{H}$ can be viewed as a maximal torus in the connected simply connected almost simple complex group G with root system $R^\vee$. The group G is the dual group in the sense of Langlands whose geometry of conjugacy classes ties up with the representation theory of the Hecke algebra $\mathbf{H}$.

The set $\mathbf{H}^\sqcap$ of equivalence classes of irreducible representations of $\mathbf{H}$ is partitioned into finite *packets* $\Pi(s,u)$ indexed by conjugacy classes of pairs (s,u) with $s \in G$ semisimple, $u \in G$ unipotent and satisfying the relation

$$sus^{-1} = u^q. \tag{3.2}$$

All members of the packet $\Pi(s,u)$ have the same central character s,and therefore are all subquotients of the principal series with spectral parameter s. Let

$$A(s,u) = Z(s,u)/Z(s,u)^0$$

be the component group of the simultaneous centralizer of s and u in G, and let $B(s,u)$ be the variety of Borel subgroups of G containing both s and u. Now the members of the packet $\Pi(s,u)$ are parametrized by the irreducible representations $\rho \in A(s,u)^\wedge$ which occur in the homology of the variety $B(s,u)$. The set of these $\rho \in A(s,u)^\wedge$ will be denoted by $A(s,u)^\wedge_{geom}$. Denote by

$$\pi(s,u,\rho) \in \Pi(s,u) \tag{3.3}$$

the irreducible representation of $\mathbf{H}$ corresponding to $\rho \in A(s,u)^\wedge_{geom}$. It can happen that $A(s,u)^\wedge_{geom}$ is strictly smaller than $A(s,u)^\wedge$.However $A(s,u)^\wedge_{geom}$ is never empty since the trivial representation $\rho = 1$ always does occur in $A(s,u)^\wedge_{geom}$. This is the classification of $\mathbf{H}^\sqcap$ as obtained by Kazhdan and Lusztig [KL]. The parametrization of the packets by pairs (s,u) as above is a special case of more general conjectures by Deligne and Langlands. The parametrization of the packet $\Pi(s,u)$ by $A(s,u)^\wedge_{geom}$ resembles the Springer classification of Weyl group representations [Sp].

Let $T(s,u)$ be a maximal torus of the reductive group $Z(s,u)$, and choose the maximal torus T of G such that both s and $T(s,u)$ lie in T. It follows from the classification of unipotent classes by Bala and Carter [Ca, Ch 5] that

$$(3.4) \qquad S = sT(s,u)$$

is a residual affine subtorus of T in the sense of Definition 2.1. Moreover the map

$$(3.5) \qquad (s,u) \mapsto (s,S)$$

is a bijection between pairs (s,u) with $s,u \in G$ semisimple and unipotent respectively satisfying (3.2) upto conjugation and pairs (s,S) with $s \in S$ and $S \subset T$ a residual affine subtorus upto action of W_0. Therefore we can write $\Pi(s,S), A(s,S), \ldots$ instead of $\Pi(s,u), A(s,u), \ldots$. The following result is due to Kazhdan and Lusztig [KL, Theorem 8.3] and (for the last statement to) Lusztig [Lu5, Proposition 9.1].

Theorem 3.1 *Suppose $q > 1$. The representation $\pi(s,S,\rho) \in \Pi(s,S)$ for $\rho \in A(s,S)^{\wedge}_{geom}$ is square integrable if and only if all members of $\Pi(s,S)$ are square integrable, and this happens precisely in case $S = \{s\}$ is a residual point. Moreover in this case the representation*

$$(3.6) \qquad \pi(s, S = \{s\}, \rho = 1)$$

is the unique antispherical representation of $\mathbf{H}$ *with central character* s.

Here $\pi \in \mathbf{H}^{\sqcap}$ is called *antispherical* if $[\pi|_{\mathbf{H}_0} : \mathrm{sign}] = 1$ with sign : $\mathbf{H}_0 \to \mathbb{C}$, $\mathrm{sign}(T_w) = (-1)^{l(w)}$. The next conjecture is in accordance with the general Langlands philosophy about formation of L-packets, and goes back in this precise form to Reeder [Re1, Conjecture 7.2].

Conjecture 3.2. Suppose $q > 1$, and let $S = \{s\}$ be a residual point of T. The formal degree (or Plancherel measure) of the square integrable representation $\pi(s,s,\rho) \in \Pi(s,s)$ is given by

$$(3.7) \qquad f\deg(\pi(s,s,\rho)) = \dim(\rho) f\deg(\pi(s,s,1)).$$

In particular all members of a square integrable packet differ in formal degree only by absolute (i.e. independent of q) constants. Of course this supports the hope expressed in Remark 2.6.

The map $T_i \mapsto -q^{-1}T_i$ extends to a unique isomorphism

$$(3.8) \qquad \mathbf{H}(W,q) \xrightarrow{\cong} \mathbf{H}(W,q^{-1})$$

preserving the adjoint and the canonical trace τ. On the level of representations this gives a bijection

$$\mathbf{H}(W,q)^{\wedge}_{antisph} \longleftrightarrow \mathbf{H}(W,q^{-1})^{\wedge}_{sph} \tag{3.9}$$

preserving canonical Plancherel measures. If ν_P^- denotes the restriction of the Plancherel measure ν_P on $\mathbf{H}(W,q)^{\wedge}$ to the antispherical unitary dual $\mathbf{H}(W,\, q)^{\wedge}_{antisph}$ then we get

$$d\nu_P^-(s,q) = d\nu_P^+(s,q^{-1}). \tag{3.10}$$

Using (2.19) we can compute (3.10) for $q > 1$ upto absolute constants, and thereby get an explicit formula for the right hand side of (3.7) upto absolute constants.

Remark 3.3. The classification of Kazhdan and Lusztig breaks down for multilabel extended affine Hecke algebras, whereas the method of Section 2 essentially should go through as mentioned in Remarks 2.5 and 2.6. It suggests that packets ought to be parametrized by pairs (s, S) upto action of W_0 with $s \in S$ and S a residual affine subtorus of T (once this notion is properly defined in the multivariable context [HO]). However the actual parametrization of a packet in geometric terms remains unclear in the multilabel context.

4 Affine Hecke algebras as intertwining algebras

In this section we will explain the role of affine Hecke algebras for the representation theory of reductive groups over a p-adic field. These results are due to Lusztig [Lu6] and Morris [Mo] generalizing earlier work of Iwahori and Matsumoto [IM] and Howlett and Lehrer [HL].

Let $\mathcal{G}$ be a tame separable locally compact unimodular group with Haar measure $\mu_{\mathcal{G}}$. Let $\mathcal{G}^{\wedge}$ denote the unitary dual of $\mathcal{G}$, and let $L(\mathcal{G})$ be the vector space of continuous complex valued functions on $\mathcal{G}$ with compact support. For $\rho \in \mathcal{G}^{\wedge}$ with representation space V_ρ the Fourier transform $\rho(\varphi)$ of $\varphi \in L(G)$ is defined by

$$\rho(\varphi) = \int_{\mathcal{G}} \varphi(x)\rho(x)d\mu_{\mathcal{G}}(x). \tag{4.1}$$

It turns out that $\rho(\varphi)$ is of trace class, and the Plancherel formula [Di] gives the existence of a unique nonnegative measure μ_P on $\mathcal{G}^\wedge$ such that

$$\varphi(e) = \int_{\mathcal{G}^\wedge} Tr_{V_\rho}(\rho(\varphi))d\mu_P(\rho) \quad \forall \varphi \in L(\mathcal{G}). \tag{4.2}$$

Suppose $K < \mathcal{G}$ is a compact open subgroup and let $\sigma \in K^\wedge$ with representation Hilbert space V_σ.

Definition 2 The Hecke algebra $\mathbf{H}(\sigma) = \mathbf{H}(\mathcal{G}, K, \sigma)$ is the space

$$\mathbf{H}(\sigma) = \{\varphi \in L(\mathcal{G}) \otimes \mathrm{End}(V_\sigma); \varphi(k_1 x k_2) = \sigma(k_1)\varphi(x)\sigma(k_2) \forall x \in \mathcal{G}, \forall k_1, k_2 \in K\}$$

viewed as an associative algebra under convolution by

$$\varphi * \psi(x) = \int_{\mathcal{G}} \varphi(xy^{-1})\psi(y)d\mu_{\mathcal{G}}(y) \quad \text{for } \varphi, \psi \in \mathbf{H}(\sigma). \tag{4.3}$$

We suppress the notation $*$ for convolution and simply write $\varphi\psi = \varphi * \psi$. The adjoint of $\varphi \in \mathbf{H}(\sigma)$ is defined by $\varphi^*(x) = \varphi(x^{-1})^*$ with the second $*$ the usual adjoint on $\mathrm{End}(V_\sigma)$. The hermitian inner product on $\mathbf{H}(\sigma)$ is defined by

$$(\varphi, \psi) = Tr_{V_\sigma}(\varphi\psi^*(e)) \quad \text{for } \varphi, \psi \in \mathbf{H}(\sigma). \tag{4.4}$$

Note that the unit element T_e of $\mathbf{H}(\sigma)$ is given by

$$T_e(x) = \begin{cases} \mathrm{vol}(K)^{-1}\sigma(x) & \text{if } x \in K \\ 0 & \text{if } x \in \mathcal{G}\backslash K. \end{cases} \tag{4.5}$$

The induced representation $\mathrm{Ind}_K^{\mathcal{G}}(\sigma)$ is defined on the vector space

$$\{\psi \in L(\mathcal{G}) \otimes V_\sigma; \psi(kx) = \sigma(k)\psi(x) \forall x \in \mathcal{G}, \forall k \in K\}. \tag{4.6}$$

Clearly $\mathbf{H}(\sigma)$ acts on (4.6) by left multiplication and $\mathcal{G}$ acts by right multiplication. Moreover these two actions commute and in fact are mutually centralizing. In turn this yields on the algebraic level a bijection

$$\mathcal{G}(\sigma)^\sqcap \ni \rho \longleftrightarrow \pi \in \mathbf{H}(\sigma)^\sqcap \tag{4.7}$$

between

$$\begin{aligned} \mathcal{G}(\sigma)^\sqcap = \{&\text{equivalence classes of admissible} \\ &\text{irreducible representations } \rho \text{ of } \mathcal{G} \text{ with } [\rho|_K : \sigma] \geq 1\} \end{aligned} \tag{4.8}$$

and

$$\mathbf{H}(\sigma)^{\sqcap} = \{\text{equivalence classes of finite dimensional irreducible representations of } \mathbf{H}(\sigma)\}. \tag{4.9}$$

On the analytic level the bijection (4.7) restricts after completion to a bijection

$$\mathcal{G}(\sigma)^{\wedge}_{temp} \ni \rho \longleftrightarrow \pi \in \mathbf{H}(\sigma)^{\wedge}_{temp} \tag{4.10}$$

for the corresponding tempered (or restricted) unitary duals.
Moreover the Plancherel measures are preserved by this correspondence in the sense that

$$d\mu_P(\rho) = d\mu_P(\pi). \tag{4.11}$$

Here by abuse of notation μ_P also denotes the Plancherel measure for $\mathbf{H}(\sigma)$. It is the unique nonnegative measure on $\mathbf{H}(\sigma)^{\wedge}$ such that

$$Tr_{V_\sigma}(\varphi(e)) = \int_{\mathbf{H}(\sigma)^{\wedge}} Tr_{V_\pi}(\pi(\varphi))d\mu_P(\pi) \quad \forall\varphi \in \mathbf{H}(\sigma). \tag{4.12}$$

Now the strategy of understanding the restriction of the Plancherel measure for $\mathcal{G}$ to $\mathcal{G}(\sigma)^{\wedge}$ from the Plancherel measure for $\mathbf{H}(\sigma)$ instead is only successful if the Hecke algebra $\mathbf{H}(\sigma)$ allows an alternative description in more elementary terms. The following situation illustrates in a nice way the power of such an approach.

Let F be a nonarchimedean local field with finite residue field $\mathbb{F}_q$. Let $\mathcal{G}$ be the group of F-rational points of a simple algebraic group defined over F. Suppose P is a parahoric subgroup of $\mathcal{G}$ with prounipotent radical U. Then the quotient M of P by U is the group of $\mathbb{F}_q$-rational points of a reductive algebraic group defined over $\mathbb{F}_q$. Suppose σ is the inflation from M to P of a cuspidal unipotent representation of M. The following result is due to Lusztig [Lu6] and Morris [Mo].

Theorem 4.1 *With the above notations we have an isomorphism of associative algebras*

$$\mathbf{H}(\mathcal{G}, P, \sigma) \cong \mathbf{H}(W, q^{\underline{k}}) \tag{4.13}$$

with $\mathbf{H}(W, q^{\underline{k}})$ *the (possibly multilabel, possibly extended) affine Hecke algebra associated with a root system* R.

The rank n of the root system R is equal to the difference of the split ranks of $\mathcal{G}$ and M. The group W is called the relative extended affine Weyl group of the pair $(\mathcal{G}, P)$. For the explicit determination of the relative root system R and the labels $\underline{k} = (k_i)$ from the pair $(\mathcal{G}, P)$ we refer to [Mo] and [Lu4].

Remark 4.3. Taking adjoints is preserved under the isomorphism (4.13). However the hermitian inner products (4.4) of $\mathbf{H}(\mathcal{G}, P, \sigma)$ and (2.3) on $\mathbf{H}(W, q^{\underline{k}})$ need not coincide. Indeed from (4.5) we get

$$(T_e, T_e) = \mathrm{vol}(P)^{-1} \dim(\sigma) = f \deg(\sigma)$$

in $\mathbf{H}(\mathcal{G}, P, \sigma)$ while

$$(T_e, T_e) = 1$$

for the canonical hermitian inner product on $\mathbf{H}(W, q^{\underline{k}})$. In turn this implies

$$\mu_P = f \deg(\sigma) \nu_P \tag{4.14}$$

with μ_P the Plancherel measure for $\mathbf{H}(\mathcal{G}, P, \sigma)$ and ν_P the canonical Plancherel measure for $\mathbf{H}(W, q^{\underline{k}})$.

5 Formation of L-packets of square integrable unipotent representations

Let $\mathcal{G}$ be the group of F-rational points of a simple algebraic group defined over the nonarchimedean local field F. Let K be a good maximal compact open subgroup of $\mathcal{G}$, and we normalize the Haar measure $\mu_{\mathcal{G}}$ on $\mathcal{G}$ by $\mathrm{vol}(K) = 1$. The following definition is due to Lusztig [Lu1].

Definition 3 An irreducible admissible representation ρ of $\mathcal{G}$ is called *unipotent* if there is a parahoric subgroup P of $\mathcal{G}$ and a unipotent cuspidal representation σ of P (or more precisely the inflation from M to P of such a representation with M the reductive quotient of P over $\mathbb{F}_q$) such that $[\rho|_P : \sigma] \geq 1$.

In particular the unipotent unitary dual $\mathcal{G}^{\wedge}_{unip}$ of $\mathcal{G}$ is given by

$$\mathcal{G}^{\wedge}_{unip} = \bigcup_{(P,\sigma)} \mathcal{G}(\sigma)^{\wedge} \tag{5.1}$$

with P a parahoric subgroup of $\mathcal{G}$, σ a unipotent cuspidal representation of P and the union taken over pairs (P, σ) upto conjugation. Restriction to the discrete series representations yields

$$\mathcal{G}^{\wedge}_{unip,ds} = \bigcup_{(P,\sigma)} \mathcal{G}(\sigma)^{\wedge}_{ds} \tag{5.2}$$

The next corollary is immediate from (4.14) keeping in mind the normalization of Haar measure on $\mathcal{G}$.

Corollary 1 *Combination of (4.10) and (4.13) yields a bijection*

$$\mathcal{G}(\sigma)^{\wedge}_{ds} \ni \rho \longleftrightarrow \pi \in \mathbf{H}(W, q^{\underline{k}})^{\wedge}_{ds} \tag{5.3}$$

and the relation between the formal degrees becomes

$$f\deg(\rho) = P_{W_K}(q) P_{W_P}(q)^{-1} \dim(\sigma) f\deg(\pi) \tag{5.4}$$

with P_{W_K} and P_{W_P} the Poincaré polynomials of the Weyl groups associated with K and P respectively. Moreover $f\deg(\pi)$ is the canonical formal degree of π as introduced in Section 2.

It follows from Theorem 3.1 that for $q > 1$ the collection

$$\mathbf{H}(W, q)^{\wedge}_{antisph,ds} \tag{5.5}$$

is parametrized by the set of residual points in T upto action of W_0. Moreover it follows from (2.19) and (3.10) how to compute their canonical formal degrees (at least upto absolute constants).

Conjecture 5.3. For multilabel extended affine Hecke algebras $\mathbf{H}(W, q^{\underline{k}})$ the collection

$$\mathbf{H}(W, q^{\underline{k}})^{\wedge}_{ds} \tag{5.6}$$

is partitioned into nonempty packets $\Pi(s, s)$ which are parametrized by residual points $s \in T$ upto the action of W_0. The parameter s is the common central character of all representations in $\Pi(s, s)$. Moreover all members of the packet $\Pi(s, s)$ have canonical formal degrees which only differ by absolute constants and which are given by (the analogous expression for multilabel Hecke algebras of) the term on the right hand side of (2.19) for $S = \{s\}$ the residual point of T.

This conjecture essentially sums up what was said before in Remarks 2.5 and 2.6. As stated there the technique of Section 2 of a contour shift and a residue computation should suffice for the proof.

Assuming the validity of this conjecture we can now discuss the formation of L-packets of square integrable unipotent representations of $\mathcal{G}$. *Two square integrable unipotent representations of $\mathcal{G}$ should lie in the same L-packet if their formal degrees only differ by absolute constants.* This is the strategy as proposed by Reeder [Re1]. We have checked that this formation of square integrable unipotent L-packets for all exceptional groups coincides with the classification of unipotent representations of $\mathcal{G}$ as obtained by Lusztig [Lu4]. In the next section we will discuss the example that $\mathcal{G}$ is of type E_8.

Remark 5.4. Let U be the prounipotent radical of the good maximal compact open subgroup K of $\mathcal{G}$. One might expect that everything said so far for unipotent representations of $\mathcal{G}$ can be generalized for U-spherical representations of $\mathcal{G}$ as well. Indeed in this more general context the results of Morris [Mo] remain valid, and therefore the computation of formal degrees of square integrable U-spherical representation of $\mathcal{G}$ can be transferred to corresponding questions about representation theory of affine Hecke algebras.

6 The example E_8

In this final section we discuss the example that the group $\mathcal{G}$ is split of type E_8. There are upto conjugation 5 parahoric subgroups P of $\mathcal{G}$ for which the associated reductive quotient M over $\mathbb{F}_q$ has a unipotent cuspidal representation σ. The pairs (P, σ) are given in Table 6.1. The unipotent cuspidal representations were classified by Lusztig [Lu2]. Throughout this section we will use his notation and results as presented in [Ca].

The packets $\Pi(s, u)$ of square integrable representations of the Hecke algebra $\mathbf{H}(W(\widetilde{E}_8), q)$ are parametrized by residual points $s \in T$ upto action of W_0. Write s in polar decomposition

$$s = s_v s_c$$

with $s_v \in T_v$ and $s_c \in T_c$. Now s is residual if an only if

$$R(s_c) = \{\alpha \in R; s_c(\theta_{\alpha^\vee}) = 1\} \subset R(E_8)$$

is a root subsystem of rank 8, and s_v is residual with respect to $R(s_c)$. The possible s_v upto action of $W_0(s_c) = \{w \in W_0; w(s_c) = s_c\}$ are in bijection with the distinguished unipotent classes in the group $Z(s_c) = \{g \in G; gs_c = s_cg\}$. In Table 6.2 we have listed the 31 possibilities by the pairs $(R(s_c), m)$ with m an index for the distinguished unipotent class in $Z(s_c)$ (ordered such that $z(s)$ in (2.14) increases with m, cf [Ca, p.174-177]). In this table we have also given the formal degrees of the corresponding representations. The multiplication of the canonical formal degree by $P_{W(E_8)}(q)$ is in accordance with (5.4). The actual calculations were performed using maple. These square integrable packets are sometimes called *generalized special*. The Steinberg representation is one of them and corresponds to $(E_8, 1)$.

On the other extreme unipotent cuspidal representations of the group $G(E_8, \mathbb{F}_q)$ lifted to K and induced upto $\mathcal{G}$ yield irreducible *supercuspidal* representations of $\mathcal{G}$. Their formal degree equals the dimension of the original unipotent cuspidal representation σ that one starts with. Comparison of [Ca, p.488] with Table 6.2 forces to which of the 31 generalized special packets these supercuspidal representations should be attached. Likewise for the intermediate cases with (P, σ) of type D_4, E_6 or E_7 the matching of formal degrees based on (5.4) forces how the various packets should be linked. The outcome has been tabulated in Table 6.3. The L-packets of unipotent square integrable representations of $\mathcal{G}$ are obtained by grouping packets in horizontal lines together.

The fact that a generalized special packet $\Pi(s, u)$ does not fill up its L-packet is connected with the fact that

$$A(s,u)^{\wedge}_{geom} \subsetneqq A(s,u)^{\wedge}.$$

For example for $(\widetilde{E}_8, E_8, 11)$ we have $A(s,u) \cong S_5$ and $A(s,u)^{\wedge} \backslash A(s,u)^{\wedge}_{geom} = \{sign\}$. The additional supercuspidal representation accounts for the missing sign representation. It is expected that the full L-packet should be parametrized by $A(s,u)^{\wedge}$.

Table 6.1. Relative root systems

W_P	σ	relative affine root system with labels
$\emptyset$	triv	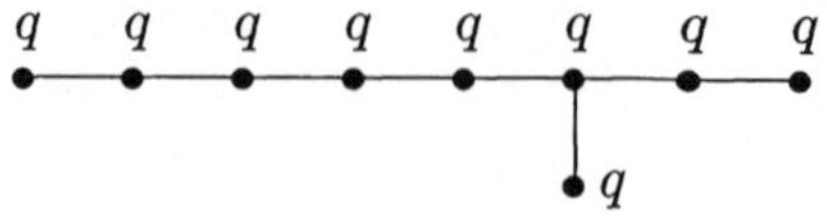
D_4	D_4	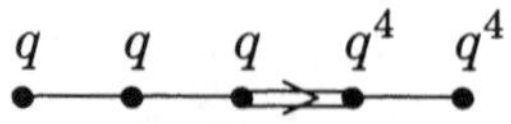
E_6	$E_6[\theta^j]$	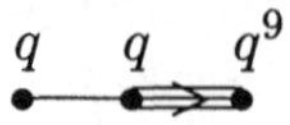
E_7	$E_7[\pm\xi]$	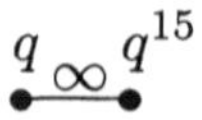
E_8	$E_8[\zeta^i], E_8[\theta^j], E_8[-\theta^j],$ $E_8[\pm i], E_8[-1], E_8^I[1], E_8^{II}[1]$	$\emptyset$

Table 6.2. Formal degrees (up to rational constants) for $H(W(\widetilde{E}_8), q)$

s	$P_{W(E_8)}(q) f \deg(\pi_s)$
A_8	$q^8\ \Phi_1^8\Phi_2^8\Phi_4^4\Phi_5^2\Phi_7\Phi_8^2\Phi_9^2\Phi_{10}^2\Phi_{14}\Phi_{18}\Phi_{20}\Phi_{27}$
$E_8, 1$	$\Phi_1^8\Phi_7\Phi_{11}\Phi_{13}\Phi_{17}\Phi_{19}\Phi_{23}\Phi_{29}$
$E_8, 2$	$q\ \Phi_1^8\Phi_7^2\Phi_9\Phi_{11}\Phi_{13}\Phi_{14}\Phi_{17}\Phi_{19}\Phi_{23}\Phi_{30}$
$E_8, 3$	$q^2\ \Phi_1^8\Phi_7^2\Phi_8^2\Phi_{11}^2\Phi_{13}\Phi_{14}\Phi_{17}\Phi_{19}\Phi_{24}\Phi_{30}$
$E_8, 4$	$q^3\ \Phi_1^8\Phi_4^4\Phi_7\Phi_8^2\Phi_9\Phi_{11}\Phi_{12}\Phi_{13}^2\Phi_{17}\Phi_{20}\Phi_{24}\Phi_{30}$
$E_8, 5$	$q^4\ \Phi_1^8\Phi_7^4\Phi_9^2\Phi_{11}^2\Phi_{13}\Phi_{14}\Phi_{18}\Phi_{20}\Phi_{24}\Phi_{30}$
$E_8, 6$	$q^5\ \Phi_1^8\Phi_5^5\Phi_7\Phi_8^2\Phi_{10}^2\Phi_{11}\Phi_{13}\Phi_{15}\Phi_{18}\Phi_{20}\Phi_{24}\Phi_{30}$
$E_8, 7$	$q^6\ \Phi_1^8\Phi_5^5\Phi_7^2\Phi_9\Phi_{10}^2\Phi_{11}\Phi_{14}\Phi_{15}\Phi_{18}\Phi_{20}\Phi_{24}\Phi_{30}$
$E_8, 8$	$q^7\ \Phi_1^8\Phi_4^4\Phi_7^4\Phi_8^2\Phi_{11}\Phi_{12}\Phi_{14}\Phi_{15}\Phi_{18}\Phi_{20}\Phi_{24}\Phi_{30}$
$E_8, 9$	$q^8\ \Phi_1^8\Phi_3^9\Phi_6^4\Phi_7\Phi_9^2\Phi_{12}^2\Phi_{14}\Phi_{15}\Phi_{18}\Phi_{20}\Phi_{24}\Phi_{30}$
$E_8, 10$	$q^{10}\ \Phi_1^8\Phi_5^5\Phi_7^2\Phi_{10}^2\Phi_{12}^2\Phi_{14}\Phi_{15}\Phi_{18}\Phi_{20}\Phi_{24}\Phi_{30}$
$E_8, 11$	$q^{16}\ \Phi_1^8\Phi_7\Phi_8^2\Phi_9\Phi_{10}^2\Phi_{12}^2\Phi_{14}\Phi_{15}\Phi_{18}\Phi_{20}\Phi_{24}\Phi_{30}$
$A_1E_7, 1$	$q^3\ \Phi_1^8\Phi_3^2\Phi_5^2\Phi_7\Phi_8^2\Phi_9\Phi_{11}\Phi_{13}\Phi_{15}\Phi_{16}\Phi_{17}\Phi_{24}\Phi_{26}$
$A_1E_7, 2$	$q^5\ \Phi_1^8\Phi_3^2\Phi_4^4\Phi_5^3\Phi_7\Phi_9\Phi_{10}^2\Phi_{11}\Phi_{12}\Phi_{13}\Phi_{15}\Phi_{20}^2\Phi_{30}$
$A_1E_7, 3$	$q^7\ \Phi_1^8\Phi_3^2\Phi_5^2\Phi_7^3\Phi_8^2\Phi_9\Phi_{11}\Phi_{14}^2\Phi_{15}\Phi_{16}\Phi_{24}\Phi_{30}$
$A_1E_7, 4$	$q^9\ \Phi_1^8\Phi_3^2\Phi_5^3\Phi_7^3\Phi_8^2\Phi_9\Phi_{10}^2\Phi_{14}\Phi_{15}\Phi_{20}\Phi_{24}\Phi_{30}$
$A_1E_7, 5$	$q^{13}\ \Phi_1^8\Phi_3^2\Phi_5^4\Phi_7\Phi_8^2\Phi_9\Phi_{10}^2\Phi_{15}\Phi_{18}\Phi_{20}\Phi_{24}\Phi_{30}$
$A_1E_7, 6$	$q^{16}\ \Phi_1^8\Phi_3^2\Phi_4^4\Phi_5^2\Phi_7\Phi_9\Phi_{12}\Phi_{14}\Phi_{15}\Phi_{18}\Phi_{20}\Phi_{24}\Phi_{30}$
$A_2E_6, 1$	$q^7\ \Phi_1^8\Phi_2^6\Phi_4^4\Phi_5^2\Phi_7^2\Phi_8^2\Phi_{10}^2\Phi_{11}\Phi_{12}\Phi_{14}\Phi_{20}\Phi_{21}\Phi_{24}$
$A_2E_6, 2$	$q^{10}\ \Phi_1^8\Phi_2^6\Phi_4^4\Phi_5^3\Phi_7^2\Phi_8^2\Phi_{10}^2\Phi_{14}\Phi_{15}^2\Phi_{20}\Phi_{30}$
$A_2E_6, 3$	$q^{16}\ \Phi_1^8\Phi_2^6\Phi_4^4\Phi_5^2\Phi_7\Phi_8^2\Phi_9\Phi_{10}^2\Phi_{14}\Phi_{20}\Phi_{24}\Phi_{30}$
$A_3D_5, 1$	$q^{12}\ \Phi_1^8\Phi_3^4\Phi_4^4\Phi_5^2\Phi_7^2\Phi_8^2\Phi_9\Phi_{12}^2\Phi_{14}\Phi_{15}\Phi_{20}\Phi_{24}$
$A_3D_5, 2$	$q^{16}\ \Phi_1^8\Phi_2^8\Phi_3^4\Phi_5^2\Phi_6^4\Phi_7\Phi_9\Phi_{10}^2\Phi_{14}\Phi_{15}\Phi_{18}\Phi_{30}$
A_4A_4	$q^{16}\ \Phi_1^8\Phi_2^8\Phi_3^4\Phi_4^4\Phi_6^4\Phi_7\Phi_8^2\Phi_9\Phi_{12}^2\Phi_{14}\Phi_{18}\Phi_{24}$
$A_1A_2A_5$	$q^{16}\ \Phi_1^8\Phi_2^6\Phi_3^2\Phi_4^4\Phi_5^2\Phi_7\Phi_8^2\Phi_9\Phi_{10}^2\Phi_{12}\Phi_{14}\Phi_{15}\Phi_{20}$
A_1A_7	$q^{11}\ \Phi_1^8\Phi_3^4\Phi_4^5\Phi_5^2\Phi_7\Phi_8^3\Phi_9\Phi_{12}^2\Phi_{15}\Phi_{16}\Phi_{20}\Phi_{24}$
$D_8, 1$	$q^4\ \Phi_1^8\Phi_3^4\Phi_5^2\Phi_7^2\Phi_9^2\Phi_{11}\Phi_{13}\Phi_{14}\Phi_{15}\Phi_{18}\Phi_{22}\Phi_{28}$
$D_8, 2$	$q^6\ \Phi_1^8\Phi_3^4\Phi_5^3\Phi_7^2\Phi_9^2\Phi_{10}^2\Phi_{11}\Phi_{14}\Phi_{15}\Phi_{20}^2\Phi_{30}$
$D_8, 3$	$q^8\ \Phi_1^8\Phi_3^7\Phi_5^2\Phi_6^4\Phi_7^2\Phi_9^2\Phi_{12}^3\Phi_{15}\Phi_{18}\Phi_{24}\Phi_{30}$
$D_8, 4$	$q^{10}\ \Phi_1^8\Phi_3^4\Phi_5^4\Phi_7^2\Phi_9\Phi_{10}^3\Phi_{14}\Phi_{15}\Phi_{20}\Phi_{24}\Phi_{30}$
$D_8, 5$	$q^{16}\ \Phi_1^8\Phi_3^4\Phi_5^2\Phi_7\Phi_8^2\Phi_9\Phi_{14}\Phi_{15}\Phi_{18}\Phi_{20}\Phi_{24}\Phi_{30}$

Table 6.3. Formation of L-packets

$W_P = \emptyset$	$W_P = W(D_4)$	$W_P = W(E_6)$	$W_P = W(E_7)$	$W_P = W(E_8)$
$(\widetilde{E}_8, A_8)$		$(\widetilde{G}_2, A_2)$		
$(\widetilde{E}_8, E_8, 1)$				
2				
$\vdots$				
10				
11				$(\emptyset, E_8^{II}[1])$
$(\widetilde{E}_8, A_1E_7, 1)$	$(\widetilde{F}_4, F_4, 1)$			
2	2			
$\vdots$	$\vdots$			
5	5			
6	6			$(\emptyset, E_8[-1])$
$(\widetilde{E}_8, A_2E_6, 1)$		$(\widetilde{G}_2, G_2, 1)$		
2		2		
3		3		$(\emptyset, E_8[\theta^j])$
$(\widetilde{E}_8, A_3D_5, 1)$	$(\widetilde{F}_4, A_1B_3, 1)$		$(\widetilde{A}_1, A_1)$	
2	2			$(\emptyset, E_8[\pm i])$
$(\widetilde{E}_8, A_4A_4)$				$(\emptyset, E_8[\zeta^i])$
$(\widetilde{E}_8, A_1A_2A_5)$	$(\widetilde{F}_4, A_2A_2)$	$(\widetilde{G}_2, A_1A_1)$		$(\emptyset, E_8[-\theta^j])$
$(\widetilde{E}_8, A_1A_7)$	$(\widetilde{F}_4, A_1A_3)$		$(\widetilde{A}_1, A_1)$	
$(\widetilde{E}_8, D_8, 1)$	$(\widetilde{F}_4, B_4, 1)$			
2	2			
3	3			
4	4			
5	5			$(\emptyset, E_8^{I}[1])$

Bibliography

[Ca] R.W. Carter, Finite groups of Lie type, Wiley, 1986.

[Di] J. Dixmier, C^*-algebras, North Holland, 1977.

[HO] G.J. Heckman and E.M. Opdam, Yang's system of particles and Hecke algebras, Annals of Math. 145 (1997), 139-173.

[HL] R. Howlett and G.I. Lehrer, Induced cuspical representations and generalized Hecke rings, Invent. Math. 58 (1980), 37-64.

[IM] N. Iwahori and H. Matsumoto, On some Bruhat decomposition and the structure of the Hecke ring of p-adic Chevalley groups, Publ. Math. IHES 25 (1965), 5-48.

[Ka] S. Kato, Irreducibility of principal series representations for Hecke algebras of affine type, J. Fac. Sci. Univ. Tokyo 28 (1982), 929-943.

[KL] D. Kazhdan and G. Lusztig, Proof of the Deligne-Langlands conjecture for Hecke algebras, Invent. Math. 87 (1987), 153-215.

[La] R.P. Langlands, On the functional equations satisfied by Eisenstein series, Lect. Notes in Math. 544, Springer Verlag, 1976.

[Le1] H. van der Lek, Extended Artin groups, Proc. Symp. Pure Math. 40 (1981), 117-122.

[Le2] H. van der Lek, The homotopy type of complex hyperplane complements, Ph.D thesis Nijmegen, 1983.

[Lu1] G. Lusztig, Some examples of square integrable representations of semisimple p-adic groups, Trans. of the AMS 277 (1983), 623-653.

[Lu2] G. Lusztig, Characters of reductive groups over a finite field, Ann. of Math. Studies 107 (1984), Princeton University Press.

[Lu3] G. Lusztig, Cuspidal local systems and graded Hecke algebras I, Publ. Math. IHES, 67 (1988), 145-202.

[Lu4] G. Lusztig, Classification of unipotent representations of simple p-adic groups, preprint MIT, 1995.

[Lu5] G. Lusztig, Cells in affine Weyl groups IV, J.Fac. Sci. Tokyo Un. 36 (1989), 297-328.

[Lu6] G. Lusztig, Intersection cohomology methods in representation theory, Proc. ICM Kyoto, 1990.

[Mac1] I.G. Macdonald, Affine Hecke algebras and orthogonal polynomials, Sém. Bourbaki 797, 1995.

[Mac2] I.G. Macdonald, Spherical functions on a group of p-adic type, Publ. Ramanujan Inst. 2, Madras, 1971.

[Mat] H. Matsumoto, Analyse Harmoniques dans les Systèmes de Tits Bornologiques de Type Affine, Lect. Notes in Math. 590, Springer Verlag, 1977.

[MW] C. Moeglin and J.-L. Waldspurger, Décomposition Spectrale et Séries d'Eisenstein, Prog. in Math. 113, Birkhäuser Verlag, 1994.

[Mo] L. Morris, Tamely ramified intertwining algebras, Invent. Math. 114 (1993), 1-54.

[Re1] M. Reeder, On the Iwahori spherical discrete series for p-adic Chevalley groups, formal degrees and L-packets, Ann. scient. ENS 27 (1994), 463-491.

[Se] A. Selberg, Discontinuous groups and harmonic analysis, Proc. ICM Stockholm (1962), 177-189.

[Sp] T.A. Springer, A construction of representations of Weyl groups, Invent. Math. 44 (1978), 279-293.

[We] H. Weyl, Über gewöhnlichen lineare Differentialgleichungen mit singulären Stellen und ihre Eigenfunktionen (2. Note), Göttinger Nachrichten (1910), 442-467.

Stability and its uses

Ehud Hrushovski
MIT
HEBREW UNIVERSITY IN JERUSALEM

The occasion for this paper is a proof of the Mordell-Lang conjecture for function fields, and associated results. Some of the results have been proved earlier by different methods, while others (notably in positive characteristic) are new. The main interest in both cases is the use of model-theoretic tools; indeed the proof requires no special ingenuity but follows naturally from model-theoretic considerations. This exposition attempts therefore to give an account of the proof and of the model theory that goes into it; I did not describe parts of model theory that are not relevant to the proof, but on the other hand made no effort to suppress model theoretic apparatus that is used in it.

§1 is devoted to the basic notions of model theory. The results mentioned there can be found in [4], though the point of view here is closer to that of Abraham Robinson. In §2 we describe parts of stability theory. For our purposes, it can be viewed as a foundations for certain kinds of geometry, analogous to algebraic geometry. In particular in §2.4-2.6 we consider a central notion, modularity, a dividing line between linear and non-linear geometries of this type. It is ultimately this dichotomy, applied to a certain structure lying inside an Abelian variety, that creates an equally sharp line between "arithmetic" Abelian varieties and others, and yields conclusions concerning non-isotrivial Abelian varieties, and torsion points of Abelian varieties over number fields.

In §2.6 we describe a classification of model-theoretic geometries having a dimension theory similar to that of algebraic geometry. §3.1 contains an application of this result to algebraic differential equations, in the form of a very rough classification of such equations into three classes (Theorem3.3). One of the classes consists essentially of those equations whose solutions can be algebraically parameterized. Another involves special differential equations, discovered by Manin, inside Abelian varieties. It may seem surprising that a very basic and general model theoretic classification, applied to the case algebraic differential equations, should lead to a consideration of Abelian varieties at all.

By a quite similar scheme we obtain similar results about difference equations, as well as certain structures in characteristic $p > 0$.

In §3.2 we deduce the geometric Mordell-Lang conjecture, and mention some other corollaries involving torsion points. Here is an oversimplified description of that proof. It concerns an Abelian variety A, a certain group Γ of points of A, and a subvariety X. We will consider $(A, +, \Gamma, X)$ as an abstract geometry of the type described in §2. We will know that the axioms of these geometries are satisfied by interpreting the situation in a differential algebraic setting (following an idea of A. Buium's), or in another setting known to have the right abstract properties. The general theory of §2 will allow us to decompose the geometry into certain "minimal" or irreducible ones. The classification in §3.1 will show that each irreducible geometry is modular, or else equivalent to algebraic geometry over a certain field k. Modularity will imply that Γ meets X in a simple way. The other alternative restricts the situation considerably and in particular, following a brief investigation of the identity of the field k, will show that A is defined over a number field.

1 Model Theory

1.1 Introduction

Model theory often deals with mathematical structures familiar from algebra or geometry, but from a point of view of its own. I will try in this section to explain this point of view.

Formally, a *structure* in the sense of model theory is a set A (or perhaps a number of sets A_ν, called *sorts*), and a collection $\mathcal{D}$ of *relations*, or *definable sets*, $R \subset A^m$ (or $R \subset A_{\nu_1} \times \ldots \times A_{\nu_m}$). One makes the assumption that $\mathcal{D}$ contains the diagonals and is closed under the following operations: intersections, complements, and inverse and direct images under projections from a product of m sorts to a partial product (such as $A^m \to A^{m-1}$.) Often $\mathcal{D}$ is generated, under these operations, by some "basic" relations $R_1, \ldots, R_k$; in this case one denotes the structure as $(A; R_1, \ldots, R_k)$, though keeping in mind the larger class of sets that can be constructed from these.

These are sometimes called *1st-order structures.* Here are some examples of structures related to number theory; (3.1-3.3) are used in the results reported on in this talk.

(1) $(\mathbb{N}; +, \cdot)$ or $(\mathbb{Z}; +, \cdot)$; $\mathcal{D}$ is generated by the graphs, in $\mathbb{Z}^3$, of addition and multiplication. This structure is the subject of Gödel's theorem (see e.g. [37]). Though one begins with the simple relations $\{+, \cdot\}$ and applies very basic geometric operations, Gödel showed that one obtains an almost arbitrary subset of $\mathbb{Z}$ within $\mathcal{D}$; the study of these sets belongs to recursion theory, or descriptive set theory, but not to geometry or number theory. Later J. Robinson showed the same is true of the field $(\mathbb{Q}; +, \cdot)$. Thus though one may ultimately be interested in $\mathbb{Z}$ or $\mathbb{Q}$, they are not directly amenable to model theoretic analysis of the geometric kind we will consider. One may note that in number theory too, one approaches these objects through a host of auxiliary structures.

(2.1) $(\mathbb{R}; +, \cdot)$. A few years after Gödel's work, Tarski showed that a quite different situation holds for the real field. The elements of $\mathcal{D}$ in this case are geometrically reasonable objects, the semi-algebraic subsets of $\mathbb{R}^n$. In particular the definable subsets of $\mathbb{R}$ itself are finite unions of intervals.

Tarski's paper was entitled "a decision procedure for elementary geometry" (and not: "for the real field"). Indeed suppose we extend this structure by adding a sort E for Euclidean 3-space, and a map $d : E^2 \to \mathbb{R}$ (distance). In this "language" we can evidently discuss triangles, circles, and general real algebraic varieties. The point is that, in the model-theoretic view of $\mathbb{R}$, this was already present. E can be interpreted as $\mathbb{R}^3$, and the relation $d(x, y) = z$ is already part of $\mathcal{D}$.

We will note below (paragraph on formulas) that if $X \subset \mathbb{R}^n$ is a definable set, the closure $cl(X)$ of X is also definable. Similarly, the derivative of a definable function is readily seen to be definable, as is distance from a point to a variety. The class of definable subsets of $\mathbb{R}^n$ thus forms a limited but useful universe; when a problem can be formulated within it, one may freely apply many geometric operations without leaving it. For some applications of this to asymptotic properties of algebraic functions, see [13], Appendix A2.

(2.2) The other (p-adic) completions of $\mathbb{Q}$, and their algebraic extensions, enjoy similar properties. It is most natural to present these fields with three sorts: K for the field itself, k for the residue field, Γ for the value field. One has relations for the valuation ring, the valuation map, and the reduction map. In particular, every subvariety of K^n is a definable set, and so is every set of the from $\{x : v(f(x)) \geq m\}$, f a rational function. The set of m'th powers in K is also definable; it is the projection to K of the graph of $y = x^m$. A theorem of Macintyre, following work of Ax and Kochen, states that all definable sets are obtained from these (using finite unions and intersections, and pullbacks by rational functions.)

(2.3) $(\mathbb{C}; +, \cdot)$, or an algebraically closed field of another characteristic. Here the definable sets are the constructible sets; finite Boolean combinations of algebraic subvarieties of $\mathbb{C}^n$ (defined over $\mathbb{Q}$.) ("Tarski-Chevalley theorem").

The examples (2.1-2.3) considered above represent a different foundational outlook on very well known geometries. Work of Denef, Van den Dries, and others, in the past decade has extended this framework to include, for example, local analytic geometry. For each type of "geometry", one must prove the analog of the Tarski-Chevalley and Tarski-Seidenberg theorems. Introducing even one new function into the class of definable sets has a high cost; one must rewrite the foundations for each such addition. (A recent example is Wilkie's addition of the global real exponential function to the local semi-analytic universe.) Once such work is done, however, one has a geometric "paradise", in which many constructions can be carried out, with a guarantee of not encountering any pathological sets on the way.

We will now turn to examples of a different kind. Classically the geometries involved are viewed as infinite-dimensional.

(3.1) A *differential field* is a field (here, of of characteristic zero) equipped with a derivation D. (One can also consider several derivations.) Together they form a structure $(U; +, \cdot, D)$. We choose $U = \mathbb{U}$ to be the *universal domain* for differential fields of characteristic zero; this notion will be explained below. An affine differential-algebraic subvariety is a subset of $\mathbb{U}^n$ defined by the vanishing of finitely many algebraic differential equations. The algebraic theory of such varieties was developed by Ritt and by Kolchin; they are also called Kolchin - closed sets. Each Kolchin closed subset of $\mathbb{U}^n$ is definable. Conversely, by a theorem of A. Robinson's, every definable set is a finite Boolean combinations of Kolchin-closed subsets of $\mathbb{U}^n$ (defined over

$\mathbb{Q}$.)

(3.2) The analog in characteristic $p > 0$ is a separably closed field $(K, +, \cdot)$, with $[K : K^p] = p$. To bring out the analogy with (3.1), one could add a derivation of K, or a stack of Hasse derivations. These are however automatically linear over K^p or K^{p^n}, hence definable whether or not they are added explicitly. (More precisely, the family of derivations forms a definable K^p-vector space, definably dual to K; a particular derivation is definable once one picks out $a \in K \setminus K^p$.) Here we will not stress this analogy, and work directly with the field structure. The definable sets will be described in §3.1.

(3.3) An analog of a different kind is obtained if one takes an automorphism in place of a derivation (difference fields). There is again a universal domain, or rather a family of universal domains depending on the restriction of the automorphism to $\tilde{\mathbb{Q}}$. Here the family of definable sets includes "difference varieties", defined by polynomials in $X, X^{\sigma}, \ldots$ where σ is a formal symbol for an automorphism. But $\mathcal{D}$ consists of something more than difference varieties; the basic definable sets describe the behavior of the automorphism on etale covers of a given variety. (See §3.1).

The examples given so far were defined in terms of a fundamental structure (the complex or real field, or differential field.) We took pains to emphasize that a whole geometry (in this case, of algebraic varieties) is generated by it. Though the geometry involves high-dimensional objects, in some sense its richness emanates from the one-dimensional fundamental structure. This duality can be reversed; when presented with a geometry synthetically, one can attempt to find an associated fundamental structure that gives rise to it. (Historically this is of course the first direction.) The next examples will be directly defined as a kind of geometry. The problem of identifying an intrinsic fundamental structure or structures responsible for their nature and complexity will be a major theme of this paper.

(4.1) Let X be a compact complex manifold; let $\mathcal{D}$ be generated by the closed analytic subvarieties of X. $\mathcal{D}$ consists in this case of finite Boolean combinations of analytic subvarieties. The fact that $\mathcal{D}$ is closed under the definable operations is due to Zilber; it relies essentially on Remmert's proper mapping theorem.

(4.2) Let A_F be the set of solutions (in the universal domain) to an algebraic differential equation $F(x, Dx, \ldots, D^n x) = 0$. If $F_1, \ldots, F_m$ are such equations, $A_i = A_{F_i}$, call a subset $W \subset A_1 \times \ldots \times A_m$ *Kolchin - closed* if it is also given as the solution set of an algebraic differential equation, in variables $X_1, \ldots, X_m$. In particular, this defines the (Kolchin)-closed subsets of $(A_F)^m$. Let $\mathcal{D}$ be the class of finite unions of differences of two Kolchin closed sets. It can be shown, using (3.1), that $(A_F, \mathcal{D})$ is a first-order structure.

If F is linear, the graph of addition on ${A_F}^3$ is a Kolchin closed set; there are many others, though they may not be immediately evident. In the non-linear case, it is less clear what $\mathcal{D}$ is, and at all events one does not expect a group or ring structure. There is no apparent intrinsic organizing principle for the class of definable sets (despite the external "upper bound" one has, in the class of all definable sets of the universal domain of differential fields). We will find such an organizing principle, arising from ideas in abstract model theory. This will in return yield information on the algebraic differential equation with which we started.

It is important not only what relations are present, but just as much so, which ones are not. For instance, compare the ring $(\mathbb{Z}, +, \cdot)$ to the group $(\mathbb{Z}, +, 1)$ in which a generator has been distinguished. Model-theoretically, they are entirely different structures. The latter has a very simple geometry of definable sets; every definable set or relation is a Boolean combination of cosets of subgroups. In particular, multiplication cannot be constructed by a finite sequence of operations of the type allowed, though it can be constructed set-theoretically since $(\mathbb{Z}, +, \cdot)$ has the same automorphism group as $(\mathbb{Z}, +, 1)$. We will see later how to establish a kind of Galois theory for this kind of question, though initially no automorphisms are present.

When seeing a copy of $\mathbb{Z}$, occurring for instance as the value group of $\mathbb{Q}_p$, the model theorist immediately asks: what kind of $\mathbb{Z}$ is it? What is the structure? (In the case of the value group of $\mathbb{Q}_p$, the answer is: precisely $(\mathbb{Z}, +, <)$; no non-obvious relations are induced on elements of $\mathbb{Z}$ as p-adic values. This is a consequence of Macintyre's theorem or Ax-Kochen-Ershov, see below.)

Formulas and sentences. Assume a structure is given in terms of some "basic" generators of the class of definable sets; for instance, the graphs of $+$ and $\cdot$ on a field. The definable sets are then formed by a sequence of geometric operations: intersection, projection, etc. A *formula* (in n variables) is a description of a definable subset (of an n-fold product, or Cartesian power), given via the procedure from obtaining it from some "basic" definable sets by projections, intersections, etc. One thinks not so much of the definable set, but of the conditions for membership in that set. Thus the intersection of two sets corresponds to the conjunction of two formulas, denoted by &. The projection of a definable set from $U \times V$ to U corresponds to the existential quantifier: for $D \subset (U \times V)$, $u \in pr_U D$ iff $(\exists v)((u,v) \in D)$. Similarly the complement corresponds to the negation symbol $\neg$.

Formally, formulas are formed by starting with a symbol for each basic definable set, including the symbol $=$ for the diagonal, as well as the symbols $\exists, \&, \neg$, and forming expressions according to certain (evident) syntactic rules. One can also use the universal quantifier $\forall$; $(\forall x)\phi(x)$ can be defined as $\neg\exists x \neg\phi(x)$; similarly "or" , "implies", etc. Any statement normally encountered in mathematics, concerning the objects at hand, can in fact by formulated with ease. However, there is a lot of weight put on the proviso "objects at hand"; one is forced to be quite precise about just what is part of the structure; in particular the collection of sequences, or even of finite subsets, of a structure A cannot directly be referred to, unless it is either explicitly put in as part of the structure or shown to be interpretable in another way.

Formulas can also be used to assert properties of the structure, rather than point to a particular definable set. By definition, a *sentence* is a formula with no free variables; it represents an theorem (or false assertion) about the geometry of the structure.

Examples of formulas.

1. in $(\mathbb{R}, +, \cdot)$, the inequality relation $\leq$- the subset of $\mathbb{R}^2$ lying above the diagonal- can be described as the projection of the affine variety $y = x + z^2$ in $\mathbb{R}^3$:

$$(\exists z)(y = x + z^2)$$

2. If $D \subset \mathbb{R}^n$ is a definable set, defined by a formula $\phi(x) = \phi(x_1, \dots, x_n)$, the closure of D is also definable:

$$(y_1, \dots, y_n) \in$$

$$cl(D) \leftrightarrow (\forall \epsilon > 0)(\exists x_1, \dots, x_n) \left(\phi(x_1, \dots, x_n) \& \sum_{i=1}^{n} (x_i - y_i)^2 < \epsilon \right)$$

3. For any field F, the following sentence is either true or false (absolutely, without reference to particular parameters):

$$(\forall a) \ldots (\forall f)(\exists x_0)(\exists x_1)(\exists x_2)(\neg(x_0 = x_1 = x_2 = 0) \\ \& \ (ax_0^2 + bx_0x_1 + cx_0x_2 + dx_1^2 + ex_1x_2 + fx_2^2 = 0))$$

It asserts that every degree 2 hypersurface in $\mathbb{P}^2$ has a solution; similarly, one can write down a sentence $\sigma_{d,n}$ stating that every degree d hypersurface in $\mathbb{P}^n$ has a solution with coordinates from the given field. The field is algebraically closed iff every such sentence holds true in it. Similarly, with infinitely many sentences, one can express the fact that a valued field is Henselian.

The linguistic interpretation of the basic operations usefully complements the geometric one. It is sometimes harder to comprehend directly the sequence of geometric operations - pullbacks, intersections, complements and projections - leading from a set to its closure, or from the graph of a function to that of its derivatives, than to read a formula such as in (2).

There are close connections between the class of sentences true in a fundamental structure, and the nature of the definable sets in the associated "geometries". Rarely does one determine one without saying a great deal about the other. For instance, it is not very different to prove the Tarski-Seidenberg theorem mentioned above, and to prove that the sentences true in $\mathbb{R}$ are those that follow from the axioms of real closed fields.

Model-theoretic categories. We can make the above discussion more explicit by describing some categories model theory works with.

Embeddings: This is the usual category used in algebra. The objects are structures whose definable sets are generated by a given collection of symbols, the *basic* relations, such as rings by $\{+, \cdot\}$. The morphisms are maps respecting a given set of relations.

Elementary embeddings : The objects are the *models* of a given *theory*, i.e. the structures in which a given collection of sentences is true. The morphisms are those embedding that respect arbitrary formulas. In other words, they respect not only the basic relations, but the comportment of the class $\mathcal{D}$ of definable sets with respect to projections, intersections, complements.

Interpretations: This is the most characteristic model theoretic category. Here the objects are structures, that need not be in the same language. The morphisms involve the linguistic viewpoint on formulas, and do not correspond to algebraic maps. An interpretation of A in B is a partial map $f : B^n \to A$, such that for any $C \in \mathcal{D}_A$, the pullback f^*C of C under f is in $\mathcal{D}_B$. In particular, the domain of f is in $\mathcal{D}_B$, and so is the equivalence relation $E = f^{-1}(=)$ on $\text{dom}(f)$.

One can therefore identify A as a set with B^n/E. The *induced structure* on B^n/E, by definition, is the class of E-invariant relations on B^n, present in $\mathcal{D}_B$. An interpretation identifies A, as a structure, with B^n/E endowed with a sub-class of the relations in the induced structure. (We will often be interested in the case where the entire induced structure is taken.)

Familiar interpretations include the "construction" of natural numbers from sets; of rational from natural numbers; of reals from the structure consisting of the rationals, and the sets of rationals; of complex from real numbers; of formulas and recursive sets in the natural numbers (Gödel); of algebraic geometric objects in the real or complex fields. Model theory undertakes to describe systematically the possible interpretations, in a given structure or class of structures.

1.2 Universal Domains

The notion of a universal domain for the class of fields was introduced by Weil in the *Foundations of Algebraic Geometry.* The idea is to have a large structure with uniform features, in which every field under consideration can be embedded; in the case of fields it will be a large algebraically closed field. Questions about a given field can sometimes be reduced to "geometric" questions about the universal domain, and "rationality" questions; for the latter one has the aid of Galois theory.

The scheme revolution put this idea into some disuse in algebraic geometry. In place of the embedding into the universal domain, one puts a structure sheaf on a variety; in this (alternate) way, one remembers the way it arose from the fundamental structure. In model theory the universal domain remains fundamental, and if a field is present, one must do some synthetic geometry to discover it. Note in particular that example (4.2) above cannot be intrinsically described as a ringed space.

Here is the general definition; it can be taken either in the embeddings or elementary embeddings category, or sometimes in intermediate ones.

Let T be a theory, i.e. a set of sentences. Let $\mathcal{C}$ be the class of structures in which T holds true: the class of fields of a given characteristic, or differential fields of char. 0, etc. One considers $\mathcal{C}$ as a category by taking morphisms to be embeddings or elementary embeddings (or sometimes an intermediate choice), as appropriate.

Definition 1.1 A structure $\mathbb{U} \in \mathcal{C}$ is a *universal domain* for T if every countable $M \in \mathcal{C}$ embeds into $\mathbb{U}$, uniquely up to an automorphism of $\mathbb{U}$.

We have combined here two properties of the universal domain; the *existence* of an embedding, together with a certain statement of existence of extensions of a given embedding, is usually called *saturation*. The *uniqueness* is called *homogeneity*.

The restriction to countable M here is somewhat arbitrary, and can be replaced with a number of variants. We have not specified $\mathbb{U}$ completely, to avoid getting into entirely irrelevant (and easily solved) set theoretic issues. For instance if the continuum hypothesis holds, one can ask that $\mathbb{U}$ have the size of the continuum; and then $\mathbb{U}$ is determined uniquely, up to a non-canonical isomorphism.

Existence. There are precise criteria for the existence of a universal domain; the essential point is that $\mathcal{C}$ should admit *amalgamation*, i.e. given two maps $f_i : A \to B_i$ there should exist $C \in \mathcal{C}$ with and $g_i : B_i \to C$ with $g_1 f_1 = g_2 f_2$. If one uses elementary embeddings, a universal domain always exists for a given (complete) theory. If one can show amalgamation for a less restrictive category, say for embeddings, one obtains additional information on the definable sets of the universal domain. (See below). This can easily be verified in our examples.

The following general procedure has been described by Abraham Robinson:

1) Identify an aspect of a given situation that you wish to concentrate on, described by universal laws; for instance the ordering on $\mathbb{Q}$, the laws being the axioms of ordered fields. Here "universal" refers to the nature of the quantifiers in the sentences describing the laws; the class of models of these sentences is to be closed under substructures.

2) Construct a universal domain for the class of structures satisfying these laws. In the case of ordered fields, one obtains a real closed field.

3) In the universal domain, projections of definable sets are usually easy to understand. If $a \in \mathbb{U}^m$ is *not* in the projection of $R \subset \mathbb{U}^{m+n}$, this cannot be due to an accidental occurrence, but rather it must follow from one of the universal laws of the class that $(a, y) \in R$ is impossible (otherwise, there would be a solution in the universal domain). This makes the class $\mathcal{D}$

comprehensible in $\mathbb{U}$.

4) We will work with the universal domain itself. However sometimes one is given in advance a particular structure, such as $\mathbb{R}$ or $\mathbb{C}$, and wishes to determine the structure of their definable sets. This can often be done most efficiently by finding first the universal domain $\mathbb{U}$ (in these examples, of ordered fields, or fields of char. 0) and then showing that the given structure forms an elementary submodel. (The fact that the definable sets of $\mathbb{U}$ are understood makes the last task easier.) In the cases of $\mathbb{C}$ and $\mathbb{R}$, Robinson noted that this procedure yields Hilbert's Nullstellensatz and Hilbert's problem on expressing rational functions as sums of squares (proved earlier by E. Artin), respectively.

The last step expresses a connection between the topological completion process, yielding a complete ordered or valued field, and the model theoretic one. The model theoretic completion "includes" the topological one, but adds also infinite and infinitesimal elements (to $\mathbb{R}$, or to the value group.)

Note that the topological option is not present in examples (3.1-3.3). The universal domain for differential fields (say) may at first seem disturbing, being an abstract object whose elements cannot be identified with (say) functions. It is a unique object, but up to a non-canonical isomorphism. This is however not essentially different than the case of the algebraic closure of a field of characteristic $p > 0$.

The key feature of the universal domain is the compactness theorem:

Theorem 1.2 (Model theoretic compactness) *Let* $\{X_i : i = 1, 2, \dots\}$ *be a family of definable subsets of* $\mathbb{U}^n$. *If every finite subfamily has non-empty intersection, then* $\cap_i X_i \neq \emptyset$

This is the fundamental finiteness principle of model theory. The apparently infinitary nature of the construction of the universal domain really goes to bring out this finiteness.

In our application, in characteristic p, we will use this theorem as follows: when A is a definable Abelian group in the universal domain, if $p^n A \neq 0$ for any n, then $\cap_n p^n A \neq 0$; moreover this intersection reflects the properties of $p^n A$ for large n.

For another application, see the Ax-Kochen theorem, discussed below.

Galois Theory. The universal domain has a large automorphism group; sufficiently large for a good Galois theory. It is also large enough for the

"Erlangen program" to hold; different "structures" have different automorphism groups. We noted above that $(\mathbb{Z}, +, \cdot)$ and $(\mathbb{Z}, +, 1)$ have the same (trivial) automorphism group, but are quite different as structures. This can no longer happen once one goes to the universal domain. If a relation R is not definable from some others $R_1, R_2, \ldots$, there exists an automorphism of the structure with respect to $R_1, R_2, \ldots$ that does not leave R invariant.

We will later need the notion of algebraic closure. If X is a subset of the universal domain $\mathbb{U}$, let $\mathrm{Aut}(\mathbb{U}/X)$ be the group of automorphisms of $\mathbb{U}$ fixing X pointwise. We define the *algebraic closure* $\mathrm{acl}(X)$ to be the set of elements $u \in \mathbb{U}$, whose orbit under this group is finite. An equivalent definition in terms of formulas: $a \in acl(X)$ iff $\phi(a, b)$ holds for some formula ϕ and parameter b from X^k, such that $\phi(x, b)$ has a finite number of solutions. The basics of Galois correspondence are valid here; in particular there is a bijection between closed subgroups of $\mathrm{Aut}(acl(X)/X)$, and subsets of $acl(X)$ containing X and closed under definable functions. (Here one must use Shelah's imaginary elements; we will not discuss this.)

Digression: The Ax-Kochen Theorem. Artin conjectured that a hypersurface of degree d in $\mathbb{P}^n$ has a p-adic solution, whenever $n \geq d^2$. This is not true ([39]), but is true asymptotically in p: for a fixed degree d, for all sufficiently large p, every hypersurface of degree d in $\mathbb{P}^{d^2}$ has a p-adic solution. Ax and Kochen ([1]) discovered and proved the asymptotic statement using the model theoretic ideas above. Recalling the proof in full would take us out of our way, but let us mention some aspects.

The *Ax-Kochen-Ershov principle* states that for a Henselian K of residual characteristic 0, the set of sentences true in K is determined by the the knowledge of those true in k (as a field) and Γ (as an ordered group.) To have an idea of the proof, recall the characterization of the Laurent series field $k((t))$ ($\mathrm{char}(k) = 0$) as the unique complete valued field, with residue field k value group $\mathbb{Z}$. Replacing "complete" by "Henselian and saturated" permits a generalization of this result to arbitrary value groups. At the level of first-order sentences - they do not change upon moving to a universal domain - it follows then that "Henselian" suffices. (cf. [27].)

Fix an integer d. We saw above that the statement of Artin's conjecture for this d (and say, for $n = d^2$) can be made by a first-order sentence $\sigma_{d,n}$. The property of being a Henselian valued field can also be expressed by an infinite collection of sentences.

Consider a structure with a sort P for the set of primes; an additional sort Q, and a function $f : Q \to P$. The fiber $Q_p = f^{-1}(p)$ of f above the prime p is taken to be the valued field $\mathbb{Q}_p$.

Embed this structure in its universal domain. The set E of elements of P such that the fiber Q_p fails to satisfy σ is a definable set. So is the set E_n of all p such that, in addition, the residue characteristic of Q_p is at least n. If the asymptotic conjecture fails, then each E_n is non-empty; by the compactness theorem , there exists an element $p^* \in P$ (in the universal domain) such that $F = Q_{p^*}$ has residue characteristic 0, and σ fails. Since each Q_p is Henselian, so is F.

It suffices now to show separately, working in characteristic zero, that σ holds in F. Ax and Kochen actually do this using compactness again: σ is known to hold (Chevalley, Tsen) for the fields $F_p((T))$; repeating the construction for these, one obtains a field F', limit of the $F_p((T))$'s. But F,F' are saturated and evidently have the same residue field and value group, hence are isomorphic by the Ax-Kochen-Ershov principle mentioned above.

2 Stability

The model theory discussed so far stayed close to algebra and its foundations. We move now to another school, pursuing model theory in the abstract. These schools developed quite separately from the the 50's to the early 80's . Moreover, starting in the 70's, the abstract model theory school begun to work with hypotheses, internally motivated, that left nearly all the significant model theoretic algebra outside the door. (The only known theories of algebraic interest satisfying the stability hypotheses are theories of modules over a fixed ring, the theory of separably closed fields, and the theory of differentially closed fields.) This isolation made possible, in laboratory conditions as it were, the discovery of concepts proper to the model theoretic view. A rapprochement began in the mid-80's, partly because of the discovery within stability theory that certain classical structures play a key role in general model theoretic phenomena, and must be explicitly studied even if one is not interested in them for their own sake. A synthesis is now in its initial stages; one hopes the power of stability-theoretic analysis will eventually become available for all of the structures surrounding number theory.

The application to Mordell-Lang reported on below uses differential algebra, in any characteristic, where stability was already applicable; and difference algebra, where the appropriate theory was developed more recently, see [22],[9].

2.1 Categoricity and dimension

The motivating problem for the development of much of stability was the "spectrum problem", the question of the number of isomorphism types of models of a given theory in a given cardinality. This takes place in the category of elementary embeddings, but the question is preserved under bi-interpretations, and encourages the discovery of invariants for the latter category.

The beginning of the subject was Morley's analysis of categoricity. In an old use of the term, a theory is called categorical if it has only one model, up to isomorphism. However, an early result (Löwenheim-Skolem, cf. [4]) states that a theory with one infinite model has models of any infinite size. Thus one says that a theory is κ- *categorical* if all its models of size κ are isomorphic. Morley showed that this notion does not depend on the cardinal κ, if $\kappa > \aleph_0$. This is a typical statement of the essentially non-set-theoretic nature of model theory. (Categoricity in the cardinality of the integers is an interesting but quite different model theoretic story, which we will leave aside.)

Note that the theory of algebraically closed fields of a given characteristic is κ - categorical, for uncountable κ: an algebraically closed field is characterized by its transcendence degree over the prime field. This transcendence degree is another face of dimension theory for varieties.

Morley started with the assumption of categoricity, and recovered an analog of transcendence degree and dimension, the Morley-rank. This is only the first of a host of dimension theories occurring in model theory, usually ordinal-valued, applying in more general situations. We will however only discuss Morley rank, and only in cases where it is finite; in this case we will refer to it as (Morley) dimension. The following definition refers to definable sets in the universal domain, possibly using parameters from the universal domain.

Definition 2.1

- A definable set X in the universal domain has Morley rank ≥ 0 iff it is nonempty.
- X has Morley rank $\geq n+1$ if for each m, it contains m disjoint definable subsets of Morley rank $\geq n$.
- If X has Morley rank $\geq n$, but not $\geq n+1$, we say it has Morley rank n. In this case there exists a maximal m, called the *multiplicity*,

such that X is the disjoint union of m definable sets of Morley rank n.

Note that X has Morley rank 0 iff it is finite. This dimension theory can be shown in general to have some of the basic properties of dimension theory for algebraic varieties; in particular, the dimension of a Cartesian product is the sum of the dimensions of the factors. (These functorial properties are as important in practice as the definition itself.)

The dimension theory serves both to begin an analysis of arbitrary definable sets, and to point out ones of particular importance.

Definition 2.2 An infinite structure U is called *minimal* if for every definable $R \subset U \times U^m$, for some integers n, n', for any $b \in U^m$:

Either $R(a, b)$ holds for at most n values of a, *or else* it holds for all but n' values of a.

If $U = \mathbb{U}$ is a universal domain, then $\mathbb{U}$ is *minimal* iff every definable subset of $\mathbb{U}$ is finite or cofinite. This includes definable subsets allowing parameters from $\mathbb{U}$.

A minimal structure has Morley dimension 1 and multiplicity 1. Conversely, every definable subset of $\mathbb{U}$ of Morley dimension and multiplicity 1 can be viewed as a minimal structure (with the induced structure.) In seeking to identify the "fundamental structures" involved in a given one, we will begin by agreeing that a "fundamental structure" should be minimal. Later we will define a notion of equivalence (non-orthogonality) to decide when two minimal definable sets correspond to the same fundamental structure. The minimal structures interpretable within a given structure M of finite Morley dimension, throw a great deal of light about M (both in abstract categoricity theory, and in the analysis and applications to be described here.)

Examples

1. Algebraically closed fields are minimal structures.

If V is any variety over an algebraically closed field K, say for simplicity without nontrivial automorphisms (as a K-variety), one can view V as a structure by taking the basic relations on V^n to be the Zariski closed subsets. Then V interprets K (hence V is bi-interpretable with the field K.)

2. Let G be a finite group, and consider the complex representations of G. The relations are the graph of addition, of the complex scalar multipliers, and of the operations of "multiplication" by any $g \in G$. There exists a universal domain for such representations (a direct sum of infinitely many

copies of the regular representation.) The Morley dimension equals the order of the group. The minimal definable sets (up to an equivalence relation to be defined below) correspond to the irreducible representations.

Within a minimal structure D, algebraic closure induces a matroid, or combinatorial pregeometry (cf. [34].) In other words, the exchange law holds: if $a \in acl(B \cup \{c\}) \setminus acl(B)$, then $c \in acl(B \cup a)$. Thus one can define a "transcendence degree" as follows: the Morley transcendence degree of a set X is the cardinality of any maximal subset Y of X, whose elements are algebraically independent.

In example (1) above, this is the usual transcendence degree. In example (2), each minimal definable set determines a different dimension theory; the transcendence degree of a representation with respect to a given minimal set, is the number of components of the associated irreducible representation that it contains.

Digression: The Baldwin-Lachlan theorem. The Baldwin-Lachlan theorem characterizes categoricity in terms of minimal structures. One says that a model M is *prime* over a subset Y if any elementary embedding of Y into $\mathbb{U}$ extends to an embedding of M into $\mathbb{U}$. In many situations, such as when M is countable, or has Morley rank, the prime model is determined up to a (non-unique) isomorphism.

Theorem 2.3 *Let T be a complete theory in a countable language. T is $\aleph_1$-categorical iff for any model M of T,*

1. *M has a definable subset D that is minimal, with the induced structure.*

2. *M is the prime model of T over D.*

Note that in (1), D is a definable subset of M, but is considered as a structure in its own right, with the structure induced from M.

The model M is thus determined by the minimal structure D. This could happen because D interprets M; an example we say earlier is that where M is a variety over an algebraically closed field F (with the "Zariski" structure). However, the situation could be considerably more complicated; a great deal of model theoretic analysis revolves around this issue. We will not directly require this analysis, though we will use some of the theory of groups in the definable category, that was developed with these issues in mind. It turns out that M can be constructed as a series of "fiber bundles", not locally trivial however, with structure groups interpretable in D.

Returning to the statement of the theorem, one sees that the minimal structure D controls in some way the behavior of the high-dimensional structure M. This fits with our theme of searching, when one is not initially known, for a fundamental structure responsible for a given geometry. Note the analogy to the Ax-Kochen-Ershov principle, stating that the simple objects - residue field, value group - in some sense control the more complex valued field.

2.2 Superstability and regular types

An extensive deepening and generalization of the Morley- Baldwin-Lachlan theory was achieved by Shelah ([36]). His main goal, irrelevant for our present purposes, was replacing the categoricity condition by any nontrivial constraint on the number of isomorphism types of models. One of the outcomes was a technology that applies to theories with ordinal Morley rank, and in particular to differential fields. The key notion is that of a *regular type*, generalizing the minimal structures occurring in the previous theory. These are orbits of the automorphism group of the universal domain (over some base structure), forming a pregeometry under a natural dependence relation. Thus there is an integer-valued dimension theory associated with every regular type. Shelah's theory shows that there are enough regular types to control the entire structure.

Much of the power of Shelah's theory resides in its ability to treat the regular types at higher tiers with nearly the same facility as the minimal ones; it is likely that future applications will utilize that power. For a start, one would like to see a classification of the regular types associated with PDE's (differential fields with several derivations). However, the present applications use only the finite rank theory (together with the ability to recognize it within an infinite rank ambient structure); we will limit out discussion to this case.

We will mention just one statement of Shelah's theory, simplified by being specialized to the finite Morley rank context. (Thus avoiding the problematics associated with one regular type embedded in another.) We first define the equivalence relation on minimal sets that we alluded to earlier.

Definition 2.4 Let D_1,D_2 be two minimal definable sets within a structure M. D_1, D_2 are *equivalent* if there exists a minimal D and definable, finite-to-one maps $f_1 : D \to D_1$, $f_2 : D \to D_2$, with cofinite ranges.

This notion of equivalence is very coarse from the point of view of algebraic geometry; with the structure of Zariski closed subsets, a variety is minimal iff it is a curve; and all curves are equivalent to each other! But indeed all curves are bi-interpretable; what they interpret, in effect, is algebraic geometry. In general the equivalence relation is slightly stronger than bi-interpretability, but is reasonable for this category. It can be useful if we are interested in determining whether a given structure belongs to algebraic geometry, or to some other domain.

We wish to point out the phenomenon of "orthogonality" of distinct dimension theories within stability theory. For minimal sets, it expresses itself as follows: if two (or more) minimal sets are inequivalent, they do not interact at all.

Lemma 2.5 (Orthogonality) *Let $D_1, \dots, D_k$ be pairwise inequivalent minimal definable sets in a universal domain $\mathbb{U}$. Then the following statements hold:*

- *Every definable $R \subset {D_1}^{m_1} \times \dots \times {D_k}^{m_k}$ is a finite union of "rectangular" sets of the form $R_1 \times \dots \times R_k$, R_i a definable subset of ${D_i}^{m_i}$.*

- *If the D_i are defined over some algebraically closed set B,* $\mathrm{Aut}(D_1 \cup \dots \cup D_k/B) = \Pi_i \mathrm{Aut}(D_i/B)$.

Both conclusions assert that the definable sets D_i are "unrelated". the 2nd statement brings out the strength of the first; it is equivalent to it, because of the homogeneity of the universal domain.

Let us note how the case $k = 2, m_1 = m_2 = 1$ of the first conclusion follows from the definitions. Remove from R the horizontal or vertical lines $a \times D_2$, $D_1 \times b$ that it may contain (or even contain up to a finite set.) Let $\pi_i : R \to D_i$ be the projections. For $a \in D_2$, ${\pi_2}^{-1}(a)$ is either finite, or a cofinite subset of D_1, by the minimality of D_1. Since we removed the lines, it must be finite. But then the maps $\pi_i : R \to D_i$ are finite, and D_1,D_2 are equivalent.

Finally, let us mention in passing how regular types enter when the object of study is the category of models and elementary embeddings. There is a notion of a finitely generated extension of a model M (prime model over M + a finite set.) There is also a notion of "tensor product" of two extensions of M. (Prime model over their independent join, see below.) This operation yields a semi-group. Shelah's theory (in the finite Morley rank context) shows that it is a free Abelian semi-group. The generators are in one-one correspondence with equivalence classes of minimal definable sets. Two such minimal sets D,D' are equivalent iff $M(a)$, $M(a')$ are isomorphic over M when $a \in D$,$a' \in D'$. If they are inequivalent, then the two extensions are "free" over M, i.e. the tensor product is their only possible relation inside a larger common extension.

2.3 Stability

The notion that lends its name to the subject turns out to be a simple combinatorial condition on bipartite graphs. It is valid in all graphs interpretable in structures with ordinal Morley rank, in particular in those occurring in algebraic or differential algebra. It has consequences of surprising elegance and force: it brings the critical notion of a "generic" element into the subject; and permits the development of a smooth and very powerful calculus of dependence and independence.

We will consider bi-partite graphs $(P, Q; R)$, or more generally pairs of sets P,Q together with relations $R_1, \ldots, R_k \subset P \times Q$.

A key example is L_n, the bipartite graph consisting of two copies P,Q of $\{1, \ldots, n\}$, and R is the graph of the order relation $i < j$. $(P, Q; R)$ embeds L_n iff there exist $a_i \in P, b_i \in Q$ $(i = 1, \ldots, n)$ such that $(a_i, b_j) \in R$ iff $i < j$.

Definition 2.6 $(P, Q; R_1, \ldots, R_k)$ is *combinatorially stable* if for each i, for some n, $(P, Q; R_i)$ does not embed L_n.

Given $R \subset (P \times Q)$, we denote

$$R(b) = \{a \in P : (a, b) \in R\}$$

for $b \in Q$; we then think of the graph as a family of subsets of P, indexed by Q. Dually, $R'(a) = \{b \in Q : (a, b) \in R\}$.

A type of example that will concern us later is called a pseudo-plane. One should think of P as a rough kind of "plane", with the $R(b)$ analogous to lines on the plane.

Definition 2.7 $(P, Q; R)$ is a *pseudo-plane* if each $R(b)$ is infinite, each $R'(a)$ is infinite, but there exists m such that for distinct $b_1 \neq b_2 \in Q$, $|R(b_1) \cap R(b_2)| \leq m$.

Sometimes one adds the dual condition that "through two distinct points there pass at most finitely many lines".

Note that all pseudo-planes are combinatorially stable.

We will make a homogeneity assumption that will make the statement of the theory easier. It is not essential; in particular if the weak primitivity condition below is omitted, one shows that there exists a definable equivalence relation with finitely many classes, such that it holds on each class.

Assumption 2.8 $\Gamma = (P, Q; R_1, \ldots, R_k)$ *is combinatorially stable. A group G acts on Γ by automorphisms. G is transitive on P. Moreover, there is no definable, G-invariant equivalence relation on P with finitely many classes.*

Theorem 2.9 *There exists a unique definable relation $\mathcal{F}_\Gamma \subset (P \times Q)$ such that:*

- *For $b \in Q$, $\mathcal{F}_\Gamma(b)$ is a intersection of sets of the form $R_i(b)$ and their complements.*
- *For any $b_1, \ldots, b_k \in Q$, $\mathcal{F}_\Gamma(b_1) \cap \ldots \cap \mathcal{F}_\Gamma(b_k) \neq \emptyset$*

Note that $\mathcal{F}_\Gamma$ is G-invariant (by the uniqueness). The relation $\mathcal{F}_\Gamma$ is called the combinatorial non-forking, or freeness relation. For instance, if $P = Q$ and R is the graph of equality, then $\mathcal{F}_\Gamma$ is the complement of R. If R is the graph of inequality, then $\mathcal{F}_\Gamma = R$. If R is a pseudo-plane, again $\mathcal{F}_\Gamma$ is the complement of R. More generally, when $k = 1$ say , $\mathcal{F}_\Gamma(b)$ is either $R(b)$ or $P \setminus R(b)$, whichever is "larger" or "more general"; $\mathcal{F}_\Gamma$ is the relation one expects between two random points. In the presence of an ordering, one expects $x < y$ or $y < x$ with equal probability, but in the stable regime the measure is 2-valued.

One has a *symmetry* phenomenon: if P, Q both satisfy 2.8, and $R^t \subset (Q \times P)$ denotes the transposed graph, $\Gamma^t = (Q, P, R^t)$, then Γ^t is stable, and $\mathcal{F}_{\Gamma^t} = {\mathcal{F}_\Gamma}^t$.

Definition 2.10 We say that $a \in P$ is *generic* over a set A, with respect to Γ, if $a \in \mathcal{F}_\Gamma(b)$ for any $b \in Q \cap A$.

In a saturated model, in particular in the universal domain, the finite intersection property in the definition of $\mathcal{F}_\Gamma$ together with the compactness theorem ensure that over any countable set A, generic elements exist. These are elements of P "in general position".

Stable structures.

Definition 2.11 A theory (or structure) is *stable* if every graph interpreted in it is combinatorially stable.

One uses this as follows. Let P and Q as two parts of a given structure, say with Q definable over a countable set A, and P forming an orbit of $Aut(\mathbb{U}/A)$. The assumption 2.8 will hold if A is *algebraically closed.* (For this one needs to take into account "imaginary elements", i.e. elements of structures interpretable in the original structure, and we will do this.) Let R_i index all the definable relations on $P \times Q$ (over A). One obtains a single "freeness" relation $\mathcal{F}$; it is the intersection of the freeness relation $\mathcal{F}_i$ corresponding to R_i. One can still use the terminology of genericity.

Definition 2.12 Let P be an orbit of $Aut(\mathbb{U}/A)$, $a \in P$, and let B be a algebraically closed substructure, $A \subset B$. Then a is *generic* over B if for any definable Q,R a is generic over B with respect to $(P, Q; R)$

It is a characteristic of stability that any two elements of P generic over B are conjugate, not only under $Aut(\mathbb{U}/A)$, but also under $Aut(\mathbb{U}/B)$. Thus given $b \in Q$, a relation $R(a, b)$ holds either for all or for no generic elements $a \in P$.

The uniqueness of the non-forking relation leads to compatibilities between the $\mathcal{F}_\Gamma$ of the various definable relations. One consequence that should be stressed in this context is Shelah's definability lemma. It has the effect that the class of definable sets is closed also under constructions using expressions such as "for a generic element $a \in P, \ldots$ ". We will see an example in 2.16 below.

Lemma 2.13 *Let R be a definable relation on $P \times Q$. Then*

$$\{b \in Q : \text{ for some generic } a \in P, R(a, b)\}$$

is a definable subset of Q.

Proof. Indeed, it is just $\{b \in Q : \mathcal{F}_\Gamma(b) = R(b)\}$, where $\Gamma = (P, Q; R)$.

Independence An equivalent language, exhibiting better the symmetry, is the language of independence.

Definition 2.14 Let A be a countable subset of $\mathbb{U}$, $a, b \in \mathbb{U}$, and let P,Q denote the orbits of a, b under $Aut(\mathbb{U}/A)$. For any definable relation R, let $\Gamma_R = (P, Q, R \cap (P \times Q))$. We say that a, b are *independent over* A if for any such R, $(a, b) \in \mathcal{F}_{\Gamma_R}$.

Then a,b are independent over A if a is a generic element of P, not only over A but also over $A \cup \{b\}$.

Two of the formal properties of independence are symmetry and transitivity: a, b are independent over A iff b, a are; if a, b are independent over $A \cup \{c\}$, and a, c are independent over A, then a,(b, c) are are independent over A.

Every structure of (finite) Morley rank is stable. Examples (2.1), (3.1), (3.2), (4.1) are stable, hence so are all structures interpretable therein; in particular, (4.2). In the case of subfields of $\mathbb{C}$, a,b are free over a subfield A iff the fields $A(a)$,$A(b)$ are free over A in the sense of field theory. (One can refine to a notion generalizing linear disjointness.)

The structures of finite rank that we will consider within difference fields (analogously to 4.2) are not stable; but enough of the bipartite graphs interpretable there are stable to explain the interactions between different part of the structure, and create a good theory of independence. The 2-valuedness referred to above no longer holds however.

Stable groups and Internality. We would like to use the above notions to say something about the possible interaction between a definable group G, and another definable set P. Generally one can show that a certain quotient of G is "internal" to P, isomorphic to a group interpretable over P; whereas the quotient is "orthogonal" to P, cannot interact with P at the level of generic elements. We will prove a special case of this here.

A definable group is a group object in the category of definable sets; i.e. a group whose underlying set and operations are all definable. We say that G is *connected* if it has no definable subgroups of finite index. The group G acts on itself by left translation; if G is connected, the hypothesis 2.8 at the beginning of this subsection holds for $P = G$, eliminating the need to work over an algebraically closed substructure.

Let P be another definable set; view it as a structure in its own right, with the induced structure.

Definition 2.15 G is *P-internal* if there exists a definable isomorphism in $\mathbb{U}$ between G and a group interpretable in P.

This stands at the opposite extreme of *orthogonality* between G and P.

We will need the following lemma, to recognize when a definable group in some part of a large structure $\mathbb{U}$, actually belongs to P.

Lemma 2.16 *Suppose G is a definable group, P a definable set, and every element of G is in the algebraic closure of some elements of P. Then there exists a finite normal subgroup K of G such that G/K is P-internal.*

Proof. We consider the Abelian case for convenience. The assumption implies the existence of a definable relation $R \subset P^n \times G$, such that $R(b)$ is finite for $b \in P^n$, and $\cup_b R(b) = G$. Let $R'(a) = \{b : a \in R(b)\}$, $K = \{b \in G : \text{ for generic } a \in G, R'(a) = R'(ab)\}$. Then K is definable, by lemma 2.13. It is easy to check that K forms a subgroup. For a fixed a, $\{b : R'(a) = R'(ab)\}$ is finite. Fix a generic element $(a_1, \dots, a_k)$ of G^k. One can show that if k is chosen large enough, then for any $b \in K$, some i is "sufficiently generic over b", so that $R'(a_i) = R'(a_i b)$. Hence K is finite. Moreover, mapping an element $b \in G$ to the k-tuple $R'(a_1 b), \dots, R'(a_k b)$ is an injective map on G/K. The range of this map consists of definable subsets of P rather than elements; however these are uniformly definable subsets, and can be parameterized by a (definable) "moduli variety", so that we get elements. We can then construct a group H in P such that the map becomes an isomorphism. □

The definition and theory of stability are due to Shelah. In the setting of groups, it is Poizat's, [33]. For a treatment along the lines described here, see [22].

2.4 Local modularity

All the structures we have considered so far possess considerable geometric complexity. We need to consider now another, much simpler class of structures, where the definable sets behave as if they were defined by linear equations. This class forced itself on the attention of model theorists in many different ways; this is reflected in its many synonymous names: 1-based, weakly normal, locally modular, linear type, (disintegrated or) module-like. There is a sharp divide between linear and nonlinear phenomena in model theory, in this sense. Moreover, certain problematics exist exclusively in the geometrically simpler "linear" case. See the survey [14] for evidence of this.

Lemma 2.17 *The following conditions on a stable structure are equivalent.*

1. *The lattice of algebraically closed subsets of $\mathbb{U}$ satisfies the modular law.*

2. *Any two algebraically closed subsets are independent over their intersection. (For this, all interpretable sorts must be taken into account.)*

3. *There is no psuedo-plane* $(P, Q; R)$ *with* P *an orbit of* $\mathrm{Aut}(\mathbb{U}/B)$ *for some finite or countable* B*, and* Q*,*R B*-definable.*

(3) asserts the non-interpretability of an even rudimentary analog of point-line geometry. (1) suggests a relation with modules; indeed there is a very close relation. Any module is stable and locally modular, if one takes the basic definable relations to be the ones defined by R-linear equations. Conversely, under certain non-degeneracy assumptions, one can show that a locally modular theory interprets an Abelian group, and behaves like a module within that group.

In the finite Morley rank context, the local modularity condition can also be stated as follows: on a definable set of Morley dimension n, there is no l-dimensional family of k - dimensional subsets, with $k+l > n$ ([6],[10]). One needs of course to define precisely the notions of a family and its dimension. Contrast this however with the case of the projective plane in algebraic geometry, of dimension 2, where the hypersurfaces of degree d form a family of 1-dimensional subsets, of dimension $(d^2 + 3d)/2$. It can be shown thatin any non-locally-modular minimal set, the plane (of dimension 2) will contain arbitrarily large familes of 1-dimensional subsets.

We note a local-global principle ([6]), demonstrating the qualitative control of minimal types over a finite-dimensional structure.

Theorem 2.18 *A structure of finite Morley rank is locally modular iff every rank-one structure interpretable therein is locally modular.*

It can be shown ([5])that the "nondegenerate" parts of a locally modular structure are controlled by definable Abelian groups. Inside these groups, the name "module-like" is justified by the following (due to Zilber in low rank and to Pillay in the $\aleph_1$-categorical context, where we will use it.)

Theorem 2.19 *[21] Let* G *be an Abelian group interpretable in a locally modular stable theory, and let* X *be a definable subset. Then* X *is a finite Boolean combination of cosets of definable subgroups.*

The conclusion of this theorem is in clear analogy with that of the Mordell-Lang conjecture (see §3). Indeed this conjecture can be viewed, model theoretically, as stating that certain diophantine-geometric structures are locally modular. The robustness of the model theoretic notion of local modularity will allow us to transfer the question from a high-dimensional Abelian variety, where the geometry is hard, to a one-dimensional object, where we can sometimes solve it.

2.5 Zilber's Trichotomy

Let us consider three types of minimal structures D.

(1) D is called *disintegrated* if the corresponding combinatorial geometry is trivial: $\mathrm{acl}(X \cup Y) = \mathrm{acl}(X) \cup \mathrm{acl}(Y)$ for $X, Y \subset D$. Equivalently, every *family* of definable subsets of D^2 is a combination of degenerate ones, in one of the following senses: it is constant (consists of a single definable subset of D^2); it is a family of points; or it is one of the familes $(D \times \{b\} : b \in D)$ or $(\{a\} \times D : a \in D)$. There is no *moving* geometry of interest on D^n, for such a D.

There are many very interesting disintegrated minimal structures, though evidently their interest does not derive from their dimension theory or algebraic closure relation. Any infinite, locally finite simplicial complex, with transitive automorphism group on vertices, forms such a structure. The basic (n-place) relations are given by the $n-1$-dimensional simplices; the other definable relations are all Boolean combinations of local ones, i.e. relations R such that for any x, there are only finitely many $y_1, \ldots, y_k$ with $R(x, y_1, \ldots y_k)$. The transcendence degree of such an object is just the number of components.

It is not known if any such (infinite, connected) disintegrated structure can be interpreted using algebraic differential equations. At all events, we will see that they cannot exist inside a finite dimensional group; therefore they will play no part in the results on Abelian varieties.

(2) D non-disintegrated, but locally modular. An example of this is a vector space V over a field or division ring F, with the definable subsets of V^n being (finite Boolean combinations of) solution sets to F-linear equations $\sum_{i=1}^{n} \alpha_i x_i = 0$. Note that there are now somewhat more moving families of "lines" in D^2, but not many; every such family is one-dimensional.

Using results mentioned in 2.4, it can be shown that up to equivalence, D can be taken to carry an Abelian group structure; and then the possibilities can be completely classified, [20]. A vector space modulo a finite subgroup is a quite typical example.

(3) An algebraically closed field, or a curve over such a field; the basic closed sets are the Zariski closed ones.

Zilber [40] conjectured that these three types are the only ones, in the sense of non-orthogonality. This can be viewed as a foundational statement, stating that algebraic geometry is sui generis; Every "geometry" with the same general dimension- theoretic properties is thus conjectured to be either at most of linear complexity, or to include algebraic geometry. A number of results in this direction exist. The conjecture itself is not true with the original set of assumptions. ([16]). It seems that the counterexamples, though not disintegrated as in type (1), have a tree-like or hyperbolic flavor. It turns out also to be possible to combine different minimal structures (of different types), without any known mechanism for taking them apart. At the moment a satisfactory substitute conjecture is missing. Nor do we know whether any counterexamples occur "in nature". The conjecture becomes true, however, under certain stronger assumptions on the dimension theory. These stronger assumptions hold in all examples mentioned in this paper. The number theoretic application we report on in §3 is based on this result.

For *real* algebraic geometry, incidentally, an analog of the conjecture was recently proved , [32].

2.6 Zariski Geometries

In the survey paper [25], Zariski geometries are motivated ab initio. Here we will view them as a topological setting for the study of minimal structures.

Definition 2.20 A *Zariski geometry* is a minimal structure D, together with a distinguished subcollection $\mathcal{C} \subset \mathcal{D}$ (called "closed sets"), and satisfying:

1. C contains the diagonals, and is closed under intersections, finite unions, and inverse images of projections.

2. $\mathcal{D}$ is generated by $\mathcal{C}$ as a Boolean algebra (on each D^n).

3. Every closed set is the union of finitely many closed, irreducible sets.

4. If X is a proper subset of Y, both closed subsets of D^n, and Y is irreducible, then $\dim(X) < \dim(Y)$

5. If X is a closed, irreducible subset of D^n, $\dim(X) = m$, and Δ is a diagonal $x_i = x_j$, then $X \cap \Delta$ is the union of closed irreducible sets of dimension at least $m - 1$.

Here "dimension" refers to the Morley rank. The first four items assert a compatibility between the topology and the definable sets, similar to the relation between closed and constructible sets in algebraic geometry. The fifth item is the new structural hypothesis. It holds in smooth algebraic varieties (and normal l.c.i varieties.) In [24], we make an effort to classify Zariski geometries as such (see the introduction there.) Here we are content with understanding them in the coarser category of minimal structures.

Theorem 2.21 *Zilber's conjecture holds for minimal structures admitting a Zariski geometry.*

In fact we will only use part of this theorem: a Zariski geometry is either locally modular, or interprets a field of Morley dimension 1.

To give a brief idea of the proof, recall a well known *linear* analog, the fundamental theorem of projective geometry. This theorem recognizes a division ring (and a finite dimensional linear geometry over it,say) in terms of intersection properties of points, lines and planes of that geometry. Of course these properties are much more complicated in the nonlinear case, when we are trying to recover a field from the intersection structure of Zariski closed sets. The solution is to linearize; in the tangent bundle we ought to find something like projective geometry. From non-local modularity, one can show the existence of families of "plane curves" (one dimensional subsets of D^2) of arbitrarily high dimension. We consider the sub-family of curves passing through a point $p \in D^2$, and wish to see its trace in an "infinitesimal neighborhood" of this point. The problem is to define the tangent space, or the equivalence relation of tangency, using synthetic geometry. Equivalently, one wants to define a limit of a family of definable sets C_{p_1,p_2} as the two points "approach each other". This can be done by considering the total family in an appropriate space, intersecting with the diagonal, removing uninteresting components, and recognizing the component corresponding to pairs of tangent curves. This uses the "dimension theorem" (property 5). In practice, however, it is difficult to define tangency well directly in high dimensions, and the proof stays close to dimension one, defining first the multiplicative and then the additive group by successive "synthetic differentiation".

2.7 Abelian groups of finite Morley dimension

Having gathered information about the minimal geometries involved in a given structure, one would like to globalize it to the full structure. A very large chapter of stability theory is devoted to this effort. One finds a sequence of fibrations of the given structure, with fibers either interpretable in minimal geometries, or else transitively acted on by groups interpreted over these geometries. Because of the possible occurence of finite groups, the situation is nontrivial even when the only minimal geometriy involved is the completely degenerate one consisting of a pure set (with no relations but equality.)

When the entire structure is contained in a finite-dimensional Abelian group, however, none of this happens, and we will be able to give a complete account of the transfer of information from the minimal geometries upward.

The first lemma is an easy generalization of facts from algebraic groups. Fix a definable Abelian group A, of finite Morley dimension. We will take the minimal subsets to be *indecomposable* (as we may), meaning they cannot be cut into finitely many pieces by the cosets of a definable subgroup of A .

Lemma 2.22 *Let X be a one-dimensional indecomposable definable subset of a finite - dimensional definable group A, $0 \in X$. Then X generates a subgroup B of A, in finitely many steps; for some n, $B = \{\sum_{i=0}^{n}(-1)^i x_i : x_i \in X\}$. Moreover, if $\dim(B) = d$, a generic element of B can be expressed in finitely many ways (at least one) as a sum of d elements of X*

Proof. The first statement, due to Zilber, generalizes the "indecomposability" theorem of algebraic groups. The proof is the same, once one has learned to replace the set - theoretic stabilizer:

$$Stab_{set}(X) = \{a \in A : a + X = X\}$$

By the more robust dimension-theoretic stabilizer:

$$Stab(X) = \{a \in A : \dim((a + X \setminus X) \cup (X \setminus (a + X))) < \dim(X)\}$$

They coincide for closed irreducible sets, but here we do not have this notion.

The second statement comes from easy dimension counting. □

The concern of the remaining lemmas is the possible mixture of orthogonal geometries. It thus has no analog in algebraic geometry proper (or in any *single* geometry, in the sense of non-orthogonality equivalence). Example (4.1) serves better. Suppose X is a complex torus; the structure is generated by the closed analytic subvarieties. If X is an Abelian variety, this is by-interpretable with the field $\mathbb{C}$; there is plenty of structure. If X is chosen generically, $\dim(X) > 1$, then X^n has no closed analytic subvarieties, except for translates of group subvarieties; X is then a minimal structure of locally modular type. (cf. [25]). (Away from the tori, we can also find disintegrated geometries in 4.1.) The next corollary implies (in the general context) that X admits a composition series into the two types. One clearly cannot expect to improve this to a direct sum decomposition. But we will do almost as well, and obtain decisive information from the "bottom layer" alone, where a direct sum decomposition does occur (2.25)

Corollary 2.23

- *A minimal subset of a group A cannot be disintegrated.*
- *Suppose A is generated by finitely many minimal subsets X_i, with each X_i equivalent to a minimal set k. Then there exists a k-internal group A' and a definable map $A \to A'$ with finite kernel.*

Proof. By 2.22, a minimal subset X of A may be taken to generate a definable subgroup B. Let (b_1, b_2) be a generic element of B^2, $b_3 = b_1 + b_2$. By the lemma, each b_i has the same algebraic closure as a d-tuple from X. Putting together the three d-tuples contradicts the triviality of algebraic dependence on X.

Similarly, if A is generated by the minimal sets X_i, then any point of A is algebraic over some points of k. The conclusion in this case follows by 2.16. □

Definition 2.24 The *socle* of A is the sum of all connected definable subgroups, generated by minimal subsets of A.

By the orthogonality lemma, the socle is a direct sum of subgroups, each generated by a finite union of minimal sets in the same non-orthogonality class. Using the known structure of locally modular groups and Corollary 2.23 above, we obtain a complete analysis of the socle.

For instance suppose A is interpretable in $\mathbb{U}_D$ or $\mathbb{U}_p$. The constant field k has precisely the structure of an (algebraically closed field) - there is no further induced structure. It follows from this that every definable group over k is definably isomorphic to a group $G(k)$, where G is an algebraic group over k. Thus the socle of A is isogenous to a direct sum of an algebraic group over k, with a locally modular group. (In the case of $\mathbb{U}_\sigma$, k is a pseudo-finite field, and the group may be a twisted algebraic group, e.g. a unitary group; this interesting variation is of no importance in the proof however.)

Call a definable group B *rigid* if every connected definable subgroup of B is defined over $\text{acl}(B)$. This sometimes follows by consideration of torsion points; it is automatically true for locally modular groups.

Proposition 2.25 (Reduction to the socle) *Let G be an Abelian group of finite Morley dimension, X an definable subset of G. Assume :*
i) The socle A of G is rigid.
ii) $Stab(X)$ is finite.
Then X is contained in a the union of finitely many cosets of A, and a definable set of smaller dimension.

3 Applications

3.1 Geometric translation

We return to examples (3) of §1.1, to discover the meaning in these concrete cases of the above concepts. Let $\mathbb{U} = \mathbb{U}_D$,$\mathbb{U}_p$,or $\mathbb{U}_\sigma$ be the universal domain of differential fields of characteristic zero, separably closed fields of imperfection index 1, and difference fields of characteristic zero (3.1-3.3, respectively.)

None of the three structures has finite Morley rank. $\mathbb{U}_D$ falls just short, with Morley rank ω; the solution set within it to any nontrivial differential polynomial does have finite Morley rank. $\mathbb{U}_p$ is stable, but has no Morley rank; nonetheless minimal types and types of finite rank exist. These "types" are countable intersections of definable sets, rather than definable sets proper; this makes little difference in the stable context, and we will ignore the distinction. $\mathbb{U}_\sigma$ is not stable at all. We will ignore this too; another dimension theory can be developed for $\mathbb{U}_\sigma$, with the same effect for our purposes ([9]).

In all three cases we are interested in the finite dimensional structures interpretable in $\mathbb{U}$. One can show ("elimination of imaginaries") that in these structures, any structure interpretable in $\mathbb{U}$ is definably isomorphic to a structure definable in $\mathbb{U}$, i.e. to a definable subset D of $\mathbb{U}^n$; the definable subsets of D^m are then among the definable subsets of $(\mathbb{U}^n)^m$. We will be interested in the maximal case, "the induced structure", where one takes the definable relations on D^m to be all subsets of D^m that are definable as subsets of $\mathbb{U}^{nm}$ (cf. §1.1).

We begin with a geometric description of the definable sets, and criteria for their finite dimensionality. Actually we will only describe a family of "basic" definable sets; but an arbitrary definable set is a finite union of differences of such basic sets. In each of the three cases, a basic definable set corresponds to an algebraic variety, together with some additional geometric data.

Basic definable sets.

$\mathbb{U}_D$ We begin by describing certain jet bundles of a variety X. We can restrict attention here to smooth affine varieties. (See [7] for a different and more detailed description. of these bundles. Note in particular that when X is not defined over the constant field $k = \{u \in \mathbb{U} : Du = 0\}$, for $m = 1$, the bundle we define is not the tangent bundle but a certain torsor thereof, determined by the structural derivation D and the Kodaira - Spencer class of X. But for our needs a more naive description suffices.)

Let $X \subset \mathbb{A}^n$ be a smooth affine variety; we will identify X with the set of points of X with coordinates in $\mathbb{U}$. When $x = (x_1, \ldots , x_m) \in X$, write Dx for $(Dx_1, \ldots , Dx_n)$. Write $D_{[m]}x$ for $(x, Dx, DDx, \ldots , D^{(m)}x)$. Let B_mX be the Zariski closure of $\{D_{[m]}x : x \in X\}$. Then B_mX is an algebraic variety; we have obvious algebraic projection maps $\pi_{m,m-1} : B_m \to B_{m-1}$. It can (easily) be shown that B_mX does not depend on the derivation D, and that B_m glues to a functor on algebraic varieties. When X is defined over the constant field, and $m = 1$, B_1X is just the tangent bundle.

A basic definable set H corresponds to an affine variety X together with a subvariety V of some B_mX, projecting dominantly to X. The set is defined by:

$$H_{X,V} = D_{[m]}^{-1}(V)$$

$\mathbb{U}_p$ In the case of char. p, one can use divided powers to give a description of the definable sets strictly analogous to the char. 0 differential case. But there is no advantage in this uniformity (for our purposes), and we will give another description.

As a field, $\mathbb{U}$ is separably closed, but has a unique purely inseparable extension $\mathbb{U}_m$ of degree $q = p^m$. The map $\phi = \phi_q : x \mapsto x^q$ is an isomorphism of of $\mathbb{U}_m$ with $\mathbb{U}$. For any variety X over $\mathbb{U}$, let X' be the variety obtained by "pulling back the coefficients using ϕ". Then ϕ_q induces a bijection $\phi_{q_*} : X'(\mathbb{U}_m) \to X(\mathbb{U})$.

On the other hand, via an inseparable analog of restriction of scalars, one can interpret the field extension $\mathbb{U}_m$ over $\mathbb{U}$. More generally, for any variety X over $\mathbb{U}$, one has a variety X_m and an isomorphism $\rho : X'(\mathbb{U}_m) \simeq X_m(\mathbb{U})$.

Composing, we obtain a bijection

$$\psi_q = \phi_{q\,*} \circ \rho^{-1} : X_m(\mathbb{U}) \to X(\mathbb{U})$$

A basic definable set (of level m) is determined by a smooth affine variety

X, and a subvariety $V \subset X[m]$. We let

$$H_{X,V} = \psi_q(V(\mathbb{U})) \subset X(\mathbb{U})$$

Here it is natural to consider also ∞-definable sets. They are determined by a compatible collection $V = (V_m)_m$, $V_m \subset X[m]$ (for each m); then $H_{X,V} = \cap_m H_{X,V_m}$. The saturation of $\mathbb{U}$ implies that this is nonempty, and Zariski dense if each H_{X,V_m} is.

The combination of compactness and stability make such ∞-definable sets no harder to handle than definable sets, and we will ignore the slight technical differences in the sequel. By a definable subset of $B = H_{X,V}$ we will mean the intersection of B with a definable set.

$\mathbb{U}_\sigma$ Here a basic definable subset of X arises from a subvariety $V \subset (Y \times Y^\sigma \times \ldots \times Y^{\sigma^l})$, where Y is a finite cover of X via a rational map $g : Y \to X$. We let $\sigma_l(x) = (x, \sigma(x), \ldots, \sigma^l(x))$ (in coordinates), and $H_{Y;V} = \sigma_l^{-1}(V)$, $H_{X,V;g} = g(H_{Y;V})$.

Criteria for finite dimension.

$\mathbb{U}_D$ If $\pi_{m,m-1}$ has finite fibers on V, then $H_{X,V}$ has finite Morley dimension, indeed dimension at most $m \dim(V)$. To see this consider the case: $m = 1$, and V is the graph of a section of B_1X. (Actually the general case can be reduced to this one by the standard trick, absorbing the higher jets into X.) In this case every basic subset of $H_{X,V}$ has the form $H_{X,V'}$, V' a subvariety of V; and V' must be of the form $V \cap B_1Y$, Y being a subvariety of X. Thus Zariski closure gives an injective map from Kolchin closed subsets of X to Zariski closed ones. The finiteness of the Morley dimension now follows from the finiteness of the usual (Zariski) dimension.

Note that not every Zariski closed subset Y of X must occur in this correspondence; e.g. when X is defined over k, only integral varieties of the vector field V do. Thus it is quite possible, indeed common, for $H_{X,V}$ to have Morley rank 1, though X has higher dimension.

$\mathbb{U}_p$ Let $V = (V_m)_m$ be data for a basic (∞-) definable set, $V_m \subset X[m]$. Then $\dim(V_1) \leq \dim(V_2) \leq \ldots$ as algebraic varieties. If the sequence of dimensions stabilizes, then $H_{X,V}$ has finite dimension. Again this dimension is induced from algebraic dimension theory on the V_m.

We will use this in the context of an algebraic Abelian group A. In this case A_m has the structure of an algebraic group, for each m. Let

L be an algebraic closure of $\mathbb{U}$; then $A_m(L) \simeq_\rho A'(L_m)$ where L_m is the commutative ring obtained from L by adding a p'th root to an element, equivalently $L' = L[\xi]/(\xi^p)$. The map $L' \to L$, $\xi \mapsto 0$ yields an algebraic group homomorphism $A_m \to A$, and it is easy to see that the kernel is a unipotent group, hence of exponent p^N for some N. Let V_m be the algebraic subgroup of $p^N A_m$ of A_m. Then the homomorphism $A_m \to A$ has finite kernel when restricted to V_m. Thus $\dim(V_m) = \dim(A)$, and by our criterion, $H_{A,V}$ has finite Morley dimension.

Under ρ, V_m corresponds to the group of elements of $A(\mathbb{U}_m)$ divisible by p^N. Thus $H_{A,V_m} = \psi_q(V_m)$ is the group of elements of $A(\mathbb{U})$ divisible by p^N, and $H_{A,V}$ is the group of infinitely p-divisible points. We obtain therefore:

The group of infinitely p-divisible elements of $A(\mathbb{U})$ has finite Morley dimension

$\mathbb{U}_\sigma$ The dimension is finite if the projection $(Y \times \ldots \times Y^{\sigma^l}) \to (Y \times \ldots \times Y^{\sigma^l})$ is finite on V.

Remarks

a) Suppose $X = H_{X,V}$ and $\dim(X) = 1$. It can be shown then that there is a fixed integer m such that for any definable $Y \subset X^n$, $Y = H_{X,V_Y}$ for some V_Y with $\dim_{\text{Zariski}}(V') = m \dim_{\text{Morley}}(Y)$. V' can be taken to be the Zariski closure of $D_l Y$, or $\lambda_l Y$, or $\sigma_l Y$ in the appropriate jet or product space. This integer need not equal 1.

b) The above remark is one reason it is helpful to use the Morley dimension theory directly, rather than (as one presumably could) translate to Zariski dimension theory in the corresponding jet or product space and do the work there. Inside a finite Morley rank X, one can pick a one-dimensional Y and work with it, as a structure. The corresponding jet space situation, is an m-dimensional object, with the curious property that the dimensions of all varieties constructed from it are divisible by m. It could be difficult to tell in advance that Y has this property, or to argue that it holds at each stage. (This is reminiscent of the situation, with $m = 2$, between real and complex geometry.)

Proposition 3.1 *Zilber's conjecture holds for $\mathbb{U}$: every minimal subset of $\mathbb{U}$ is locally modular, or interprets a one-dimensional field*

Proof. In the case of $\mathbb{U}_D$ and $\mathbb{U}_p$, call the basic sets defined above *closed.* After possibly removing a singular locus, the jet-space varieties they correspond to can be taken to be non-singular. The dimension theorem of algebraic geometry, for these jet-space varieties, can be shown to yield the dimension theorem for the minimal set. ([23] for $\mathbb{U}_D$, [19] for minimal sets in $\mathbb{U}_p$ satisfying the above finite dimensionality criterion; [11] for other minimal sets.) One then uses the theorem on Zariski geometries.

In the case of $\mathbb{U}_\sigma$, it is first necessary to show that a non-locally modular is *stable*. ([9]). At this point it may be possible to prove the dimension theorem for an appropriate Zariski geometry. In fact this course is not taken; instead a version of the following principle is used ([17]): if certain automorphisms groups of a minimal structure are unimodular as locally compact groups (equal left and right Haar measure), the structure must be locally modular. (See [17] and [9] for details.) □

Let k be the constant field of $\mathbb{U}_D$, or the fixed field of $\mathbb{U}_\sigma$, or $k = \cap_r \mathbb{U}^{p^r}$ in the case of $\mathbb{U}_p$. In all cases, k can easily be shown to be minimal; it is certainly not locally modular, since it carries a field structure.

Theorem 3.2 *Every non-locally modular minimal set D in $\mathbb{U}$ is equivalent to k (in the sense of non-orthogonality).*

Proof. By the above proposition, D is equivalent to *some* one-dimensional field k'. It is a matter of showing that all interpretable one-dimensional fields are isomorphic. This can be viewed as a special case, or at least an allied problem, to that of classifying the interpretable simple groups. Model theoretic methods are quite useful here, and for instance show effortlessly that definable groups embed in algebraic groups, with information on the embedding. One must then rely on the theory of algebraic groups. The results are due to P. Cassidy (by other methods) for $\mathbb{U}_D$, and to Ph.D. theses by Sokolovic [38] and Mesmer [29] in the cases of $\mathbb{U}_D$ and $\mathbb{U}_p$.

In the case of $\mathbb{U}_\sigma$, the definable groups and simple groups can also be classified. Twisted groups do occur; in fact the list of definable simple groups is the same as that of non-sporadic finite simple groups, and thus includes for instance the unitary groups, and in characteristic 2 analogs of the Suzuki groups. In characteristic $p > 0$, twisted *fields* occur too - such as the fixed field of $x^p\sigma$. But in characteristic 0, it can be shown that the only definable field is the fixed field k of σ, and the finite extensions thereof. □

Interesting algebraic information is obtained upon translating the above back to language of differential equations, or difference equations. We consider equations of the form

$$f(X, DX, D^2X, \ldots, D^nX) = 0 \quad \text{or} \quad f(X, \sigma X, \ldots, \sigma^n X) = 0,$$

where D and σ are symbols for a derivation, or automorphism. Let us call a differential equation *geometrically degenerate* if the set of solutions, in the universal domain, forms a minimal set of disintegrated type (in the Zil'ber classification.)

Theorem 3.3 *Any algebraic differential equation can be reduced, by a differential-algebraic change of variable, to one of the following form:*

1. *A geometrically degenerate differential equation.*
2. *An equation defining a minimal subgroup of a simple, non-isotrivial Abelian variety. (Manin's equation).*
3. *The equation $Dx = 0$ defining the field of constants.*

Remarks

1. The change-of-variable is to be performed by a differential-algebraic polynomial , that may itself involve non-constant coefficients. Thus the theorem says nothing about linear equations; there is always a differential rational function (involving a basis of solutions as coefficients) reducing an arbitrary solution to a constant. There does exist a coefficient-free theory; in this case it amounts to a slight generalization of the Lie-Kolchin Galois theory of differential equations. (Poizat, Pillay).

2. In degree one (n=1), the description of geometrically degenerate equations can be sharpened. It can be shown, using methods of Jouanolou, that such an equation is equivalent to one with no structure at all; no two solutions are algebraically dependent, and only finitely many are algebraic over any given function field of finite transcendence degree. It is possible that the same holds in higher degrees; this may have applications to rational points over function fields, of varieties of arbitrary dimension. The proof presumably requires geometric methods; but the model theory expressed in the following theorem should give a guide, if only to suggest the correct statement of the theorem in higher dimensions.

3. Any non-isotrivial Abelian variety has a unique minimal infinite subgroup definable by an algebraic differential equation. It is minimal not only as a definable subgroup, but also as a definable set. Two such groups are equivalent iff their abelian varieties are isogenous; otherwise orthogonality holds. These facts can be proved purely model-theoretically (and indeed to my knowledge no other proof exists.) The equations were found by Manin;

see Manin's paper as well as [8] for a description. The theorem shows that they have a priori significance among all algebraic differential equations, and must be taken into account in a general theory.

4. An analogous theorem is available for difference equations. Alternative (1) looks the same; (3) refers to the fixed field; (2) is similar but here Abelian varieties over number fields show up in place of non-isotrivial ones. See [9], [17].

3.2 Points on subvarieties of Abelian varieties

The Mordell-Lang conjecture grew from Mordell's 1922 hypothesis on the finiteness of the number of rational points on a curve of high genus (Falting's theorem, 1983.) It has had a complicated history, well told in a number of sources; see for instance [12], [26]. Briefly, according to the Mordell-Weil theorem, the group of rational points of an Abelian variety is finitely generated. Since a curve may be embedded in its Jacobian, Mordell's conjecture becomes equivalent to the following statement: a curve of genus ≥ 2 over a number field intersects a finitely generated subgroup of its Jacobian in a finite set.

Mordell's conjecture was transposed to function fields by Lang in 1960, and in this version was proved by Manin (in characteristic 0) in 1963 (followed by proofs by Grauert, and in positive characteristic by Samuel.) A related conjecture arose, by Manin and Mumford: a curve of genus ≥ 2 intersects the set of torsion points of its Jacobian in a finite set. (Proved by Raynaud, 1982).

This led Lang (1965) to a more general conjecture, replacing the Jacobian by an Abelian variety, the curve by an arbitrary subvariety, and the finitely generated, or torsion, group by the group of division points of a finitely generated Γ; i.e. the points a such that $ma \in \Gamma$ for some positive integer m. This group will be denoted by $\tilde{\Gamma}$. In positive characteristic p, however, following [2], we let $\tilde{\Gamma} = \{g : mx \in \Gamma_0, \text{ some } m \text{ prime to } p\}$. (The conjecture remains open for other m.)

Allowing a higher dimensional variety X in the hypothesis forces a modification in the conclusion: the intersection can no longer expected to be finite, but is rather conjectured to be contained in a finite union of group subvarieties, contained in X.

A further generalization was made to semi-Abelian varieties. See [26], 6.3. Our methods prove immediately the strongest "relative" version of the conjecture, for semi-Abelian varieties. For simplicity we will state only a weaker version and only for Abelian varieties.

Theorem 3.4 (Geometric Mordell-Lang) *Let A be a Abelian variety over an algebraically closed field K. Suppose A has no nontrivial homomorphic images defined over a finite field, or a number field. Let Γ be a finitely generated subgroup of $A(K)$, and let X be a subvariety of A. Then the Zariski closure of $X \cap \tilde{\Gamma}$ is a finite union of cosets of group subvarieties of A.*

In characteristic $p > 0$, the restriction against homomorphic images defined over finite extensions of the prime field is necessary. In characteristic 0, grace to Faltings and Vojta, it is not, and rather reflects a weakness of our present tools. We are presently unable to deal with the rational field, or with number fields of small degree. For function fields of characteristic 0, a number of proofs are available, including one by Buium ([8]) using differential algebra; Buium's proof, while quite different from ours beyond the first step, uses the Manin map in the same way, and it was seeing this that inspired the model theoretic proof. In characteristic $p > 0$, the model theoretic proof is the only one known in general; but many cases are proved in [2].

While unable to handle the general case over number fields, we can prove the Manin-Mumford conjecture; the proof has an explicit and uniform character, and appears to be the only proof to give an effective bound. (Though Hindry gives explicit bounds modulo one constant.) The proof includes the semi-Abelian case, as in [28] (according to McQuillan, with some improvements.)

For instance, we obtain:

Theorem 3.5 *Let A be a commutative algebraic group, defined over some number field, and fix a very ample divisor H on A. Let X be a subvariety, defined over K^{alg}. Then there exist a finite number M of subvarieties $c_i + A_i$ of X, all translates of group subvarieties A_i of A, such that*

$$T(A) \cap X = \cup_i T(A_i) + c_i.$$

We have $M \leq c \deg_H(X)^e$, with c and e depending on A but not on X.

c and e can also be written down explicitly; they are doubly exponential in some natural parameters associated with A.

The method immediately yields a proof of the following theorem. It is a restricted version of a conjecture due to Silverman, Tate and Voloch. Let $\mathbb{C}_p$

be the completion of the algebraic closure of the p-adics $\mathbb{Q}_p$. The conjecture concerns Abelian and semi-Abelian varieties over $\mathbb{C}_p$, but we are only able to handle finite extensions of $\mathbb{Q}_p$ therein.

Theorem 3.6 *Let A be a semi- Abelian variety defined over $\mathbb{Q}_p$, or a finite extension of $\mathbb{Q}_p$, with good reduction. Let X be a subvariety of A, defined over $\mathbb{C}_p$. Then there exists a bound $\lambda > 0$ such that for any prime-to-p torsion point a of $A(\mathbb{C}_p)$, either $a \in X(\mathbb{C}_p)$, or the p-adic distance from a to X is $\geq \lambda$.*

Let us now discuss the proof of the above theorems. For definiteness we will concentrate on 3.4 in characteristic $p > 0$, where a separably closed field $\mathbb{U}_p$ is used. The characteristic 0 case uses is so close that one could easily formulate the two cases in a common language; it uses a differential field $\mathbb{U}_D$. The proofs of 3.5 and 3.6 have a similar structure but use $\mathbb{U}_\sigma$ instead.

Recall that k is the derivation in characteristic 0; it is $\cap_n K^{p^n}$ in characteristic p; and it is the fixed field of the automorphism, in the case of U_σ.

In all cases, one replaces the finitely generated group Γ (or the group of torsion points, in 3.5) by a larger group B, defined using the derivation or automorphism or inseparable structure. One shows that if the conclusion fails, then B is internal to k.

In the case of 3.4, one chooses k in advance to be linearly disjoint from the field of definition of A over the algebraic closure of the prime field. Then the isotriviality of B yields isotriviality of A, i.e. A will have homomorphic images defined over the algebraic closure of the prime field, contradicting the assumption.

In the case of 3.5, we are already over a number field, and "isotriviality" takes the following sense. There one uses an automorphism σ, with A defined over the fixed field k, and $B = \ker F(\sigma)$ for a certain polynomial F. The fact that B is internal to k turns out to mean that F is cyclotomic. However, one chooses σ to be a lifting of Frobenius, and F to be the characteristic polynomial of Frobenius on A; and it follows that F has no cyclotomic factors.

Having mentioned these differences, we restrict attention to characteristic $p > 0$.

Proof of 3.4. We can put ourselves in the following situation: $\mathbb{U}$ is the universal domain for separably closed fields of characteristic p, with $[\mathbb{U} : \mathbb{U}^p] = p$. $k = \cap_n K^{p^n}$ is the largest algebraically closed subfield. A is a semi-Abelian variety defined over K; for simplicity, suppose A is an Abelian variety, with no nontrivial homomorphic factors defined over k. Γ is a finitely generated subgroup of $A(K)$, or the group of (prime-to-p)- division points of such a group. X is a K-subvariety of A. We want to show that the Zariski closure of $\Gamma \cap X$ is a finite union of cosets of group subvarieties of A.

Step 1. Let $B = p^{\infty}A(K) = \cap_r p^r A(K)$. Then B has finite Morley dimension. Moreover the minimal types within it satisfy Zilber's conjecture.
Proof. This was proved in §3.1.

Step 2. The minimal definable sets within B are locally modular.
Proof. By 3.2, if a non-locally-modular minimal set C occurs in B, it must be equivalent to k in the sense of non-orthogonality. Let C' be a group generated by C, as in 2.22. By 2.16, there exists an isogeny from C' to a to a definable group $H(k)$ over k. Taking essentially the Zariski closure of the isogeny, we obtain an isogeny from the Zariski closure A' of C' in A, to $H(K)$. Since H is defined over k, this contradicts our assumption on isotrivial factors.

Step 3. $X \cap socle(B)$ is a finite Boolean combination of cosets of definable subgroups of B.
Proof. Each minimal type in the socle is locally modular, and hence the socle is locally modular (2.18). Thus by 2.19, every definable subset of the socle is a finite Boolean combination of cosets definable subgroups of B.

Step 4. $X \cap B = Y \cap B$, where Y is a finite union of group subvarieties of A, contained in X. *Proof.* We can reduce to the case that the stabilizer of X is finite. Then by 2.25, X is contained in finitely many cosets of the socle of B. By Step (3), $X \cap B$ is a finite Boolean combination of cosets of (definable) subgroups of B. Let Y be the Zariski closure of $X \cap B$. One sees easily that Y is a finite union of group subvarieties.

Step 5. For some r, $X \cap p^r A(K) = Y \cap p^r A(K)$.
Proof. This follows by compactness from step (4): the intersection of all the groups $p^r A$ and of the definable set $X \setminus Y$ is empty, by (4); hence some finite intersection is empty.

In the same way, for any translate $X + b$ of X, there exists r such that the Zariski closure of $p^r A(K) \cap (X + b)$ is a finite union of translates of group subvarieties of A. Further, by compactness (1.2), one can choose the same r to work for all translates.

Now $\Gamma/p^r\Gamma$ is a finite group; so Γ is a union of finitely many translates of $\Gamma \cap p^r A(K)$, and hence lies in a finite union of cosets of $p^r A(K)$. The Zariski closure of the intersection of X with each of these cosets is a finite union of group subvarieties; hence so is the intersection of X with their union. Thus $X \cap \Gamma$ is contained in a finite union of group subvarieties contained in X, as promised. □

1. For a fixed Abelian variety A, and X varying within a bounded algebraic family of subvarieties of A, the bound is uniform. Further, since Γ only enters the proof in the last step, via the number of cosets of $p^r A(K)$ it meets, the bound on the number of cosets of group subvarieties required has the form $cp^{r\gamma}$, where c and r do not depend on Γ, and γ is the "rank" of Γ, i.e. the maximal integer such that Γ contains a copy of $\mathbb{Z}^\gamma$.

2. In characteristic 0, (3.4 or 3.5), B is definable by a single formula (rather than infinitely many), and so the use of compactness in Step (5) above is not needed. As a consequence, one gets explicit (doubly exponential) bounds.

Bibliography

[1] J. Ax and S. Kochen, Amer. J. Math 87 (1965) 605-630, 631-648, Diophantine Problems over local fields I,II; Annals of Math. **83**, pp. 437 -456, Diophantine Problems over local fields III.

[2] Abramovich, D. and Voloch, J.F., Toward a proof of the Mordell-Lang conjecture in characteristic p, Duke Mathematical Journal 1992 No. 5 pp. 103-115.

[3] L. Henkin et al, eds., Proceedings of the Tarski Symposium, Proceedings of symposia in pure mathematics XXV, AMS, Providence 1974

[4] Chang, and Keisler, J., *Model Theory* , 3rd Edition, North -Holland, Amsterdam - Tokyo 1990

[5] , E. Bouscaren and E. Hrushovski, Interpreting groups in one-based theories *Journal of Symbolic Logic,* to appear

[6] S. Buechler, Locally modular theories of finite rank, Annals of Pure and Applied Logic **30** (1986) no. 1, 83-94

[7] A. Buium, Differential Function Fields and Modulii of Algebraic Varieties, LNM 1226, Springer 1986

[8] A. Buium, Intersections in jet spaces and a conjecture of S. Lang, Annals of Mathematics 136(1992), pp. 557-567.

[9] Z. Chatzidakis and E. Hrushovski, The model theory of difference equations, to appear.

[10] G. Cherlin, L. Harrington, and A. Lachlan, $\aleph_0$-categorical, $\aleph_0$-stable structures, *Ann. Pure Appl. Logic* **28** (1985) 103-135.

[11] F. Delon, work to appear on minimal types in separably closed fields.

[12] M. Hindry Sur les conjectures de Mordell et Lang (d'apres Vojta, Faltings et Bombieri), *Asterisque* 209 (1992)

[13] Lars Hörmander, *The Analysis of Linear Partial Differential Operators II*, Springer-Verlag Berlin - Heidelberg - New - York Tokyo 1983

[14] E. Hrushovski, Unidimensional Theories: an introduction to geometric stability theory, in *Logic Colloquium 88*, Granada, North Holland 1989.

[15] E. Hrushovski, Locally Modular Regular Types, *Classification Theory: Chicago 1985*, editor J.T. Baldwin, Springer Verlag 1987

[16] E. Hrushovski, A New Strongly Minimal Set,*Annals of Pure and Applied Logic***62** (1993) pp. 147-166.

[17] E. Hrushovski, the Manin-Mumford conjecture and the model theory of difference fields, preprint

[18] E. Hrushovski, unimodular minimal structures, London Math. Soc. Journal

[19] E. Hrushovski, E., The Mordell-Lang Conjecture for Function Fields, to appear in *Journal of the AMS*

[20] E. Hrushovski, J. Loveys, A structure theorem for strongly minimal sets, preprint

[21] E. Hrushovski and A. Pillay, Weakly Normal Groups, in *Logic Colloquium 85*, North Holland, Amsterdam 1986

[22] Hrushovski, E. & Pillay, A., Groups Definable in Local Fields and Pseudo-Finite Fields, *Israel Journal of Mathematics* **85** (1994) pp. 203-262

[23] Hrushovski, E. & Sokolovic, Z., Minimal subsets of differentially closed fields, to appear in *Transactions of the American Mathematical Society*

[24] E. Hrushovski and B. Zilber, Zariski Geometries, to appear in *Journal of the AMS*

[25] E. Hrushovski and B. Zilber, Zariski Geometries, AMS *Bulletin***28** (1993) no. 2 pp. 315-323

[26] S. Lang, *Number Theory III: Diophantine Geometry*, EMS 60, Springer-Verlag, Berlin-Heidelberg 1991

[27] Angus Macintyre, 20 years of p-adic model theory, Logic Coll. 84, pp. 121-153, N. Holland Amsterdam - NY 1986

[28] Michael McQuillan, Division points on semi-abelian varieties, *Inventiones mathematicae* 120 143-149 (1995)

[29] M. Mesmer, Groups and fields interpretable in separably closed fields, Ph.D. thesis, University of Illinois at Chicago, 1983

[30] Anand Pillay, Introduction to Stability Theory, Clarendon Press, Oxford 1983

[31] Anand Pillay, Geometric Stability Theory (book, forthcoming).

[32] J. Peterzil, S. Starchenko, preprint

[33] B. Poizat, Sous-groupes définissables d'un groupe stable, J. Symbolic Logic 46 (1981) 137-146.

[34] Henry H. Crapo and Gian Carlo Rota, On the Foundations of Combinatorial Theory: Combinatorial Geometries, MIT Press, Cambridge 1970

[35] Jean-Pierre Serre, *Lectures on the Mordell-Weil Theorem*, Aspects of Mathematics E 15, Vieweg, Braunschweig/Wiesbaden 1990.

[36] Saharon Shelah, *Classification Theory*, revised edition, North-Holland Amsterdam-Tokyo 1990

[37] C. Szmoryński, Logical Number Theory I, An Introduction, Springer Verlag Berlin-Heidelberg 1991

[38] Z. Sokolovic, Model Theory of Differential Fields, PhD Thesis, Notre Dame, July 1992

[39] C.R. Acad. Sci. A-B 262 (1966)

[40] B.I. Zilber, Strongly minimal countably categorical theories, Siberian Mathematical Journal 21 (1980), 219-230, II-III (1984), pp. 396-412, 559-571

Wavelets, paraproducts, and Navier-Stokes equations

Yves Meyer
UNIVERSITY OF PARIS IX,
CEREMADE,
75775 PARIS, CEDEX 16,
FRANCE

Foreword

These notes are obviously not the written version of the two talks I gave on April 20, 1996, in the Harvard-MIT seminar.

These two talks were directed to a broad audience and I wanted to present a large panorama of applications of wavelets to science and technology.

In contrast here I wished to focus on one specific application of wavelet analysis. I decided to address a hot issue : the relevance of wavelets for Navier-Stokes equations.

These notes will lead the reader to much unpublished material on Navier-Stokes equations, and the debate will not be settled.

I would like to thank the organizers of the Harvard-MIT seminar for their warm hospitality and David Jerison for his patience and kindness.

Last but not least, Isabelle Paisant did a splendid job in typing these notes.

Yves Meyer
La Barre de Monts,
August 25 1997

Wavelets, paraproducts, and Navier-Stokes equations

1 Introduction

The relevance of wavelet analysis in Navier-Stokes equations is a serious issue. Big claims were not followed by any serious progress. The first goal of these notes is to clarify these matters.

The second goal is to pave the way to some striking results obtained by P.G. Lemarié-Rieusset and his group. On the way the fundamental achievements by T. Kato will be revisited.

All these results are concerning the same problem, namely, the existence and uniqueness of solutions of Navier-Stokes equations when the initial condition $u_0(x)$ belongs to a given functional space E. We then want to compute the velocity field $u(x,t)$ for $t \geq 0$ and to prove estimates of the form

$$\sup_{t\geq 0} \|u(\cdot,t)\|_E \leq C\|u_0\|_E \,. \tag{1.1}$$

This problem was treated by Jean Leray when $E = L^2(\mathbb{R}^3)$. Indeed Leray proved the existence of weak solutions to Navier-Stokes equations (section 4). The uniqueness of such solutions for a given $u_0(x)$ is still an open problem. The same problem was treated by T. Kato in 1984 when $E = L^3(\mathbb{R}^3)$. The existence of a global mild solution was obtained by a beautiful argument which will be unveiled in section 19. However the uniqueness problem was open until now and recently solved by P.G. Lemarié-Rieusset. This issue concerns solutions to Navier-Stokes equations $u(x,t)$, $x \in \mathbb{R}^3$, $t \geq 0$, which are continuous in the time variable with values in $L^3(\mathbb{R}^3)$ as functions of x.

On the way the relevance of wavelet analysis for Navier-Stokes equations will be carefully studied and the results are more subtle than expected.

2 Notations

Navier-Stokes equations describe the motion of incompressible and homogeneous fluids. It is natural to assume that this fluid is contained in a bounded domain Ω. However we will restrict our attention to the opposite situation where there is no boundary. In other words the fluid is filling the space $\mathbb{R}^3$.

The velocity at a given point $x \in \mathbb{R}^3$ and at a given time $t \geq 0$ of the fluid is denoted by $u(x,t) = \big(u_1(x,t), u_2(x,t), u_3(x,t)\big)$ where u_1, u_2 and u_3 are real valued.

Similarly the pressure is denoted by $p(x,t)$ and is a real valued function.

In the absence of external forces, the Navier-Stokes equations read

$$(2.1) \qquad \begin{cases} \dfrac{\partial u}{\partial t} = \nu\Delta u - \big(u_1\partial_1 + u_2\partial_2 + u_3\partial_3\big)u - \nabla p \\ \partial_1 u_1 + \partial_2 u_2 + \partial_3 u_3 = 0 \\ u(x,0) = u_0(x)\,. \end{cases}$$

Here and in what follows, $\partial_j = \partial/\partial x_j$, $\Delta = \partial_1^2 + \partial_2^2 + \partial_3^2$, ∇p is the gradient of the pressure and $u_0(x) = \big(u_{0,1}(x), u_{0,2}(x), u_{0,3}(x)\big)$ is the initial condition.

The viscosity ν is a positive parameter which can be eliminated by a convenient rescaling (section 5).

The system (2.1) contains four unknown functions u_1, u_2, u_3 and p but consists of four equations. The balance is correct.

The pressure does not show up in the initial data. It will soon be explained how the pressure can be eliminated from (2.1). Therefore (2.1) will reduce into a system of three equations governing the evolution of the velocity field.

We say that $u(\cdot,t)$ is a classical solution of (2.1) if the regularity of u with respect to x and t and the decay at infinity permit all the calculations which will now be performed.

We say that a solution $u(\cdot,t)$ of (2.1) is global in time if it is defined for $t \in [0,\infty)$ with some other required properties. These properties concern smoothness and decay at infinity. A solution $u(\cdot,t)$ of (2.1) is local in time if it is defined for $0 \le t \le T$ for some positive T.

In these notes $u(x,t)$ will be studied as a vector-valued function of the time variable. More precisely we will start with a functional Banach space $E \subset \mathcal{S}'(\mathbb{R}^3)$. This Banach space is used for describing size or regularity properties of functions $f(x)$ of the x variable. Then $u(x,t)$ will always be viewed as an E-valued function of the time variable.

Indeed we will be looking for solutions $u(x,t)$ belonging to $C([0,\infty);E)$. Here and in what follows we incorrectly write $u(\cdot,t) \in E$ instead of $u(\cdot,t) \in E^3$.

A solution u in $C([0,\infty);E)$ is certainly not a classical solution. Smoothness in t or x is missing. This issue will be addressed in section 9 where the concept of a mild solution will be defined.

Before defining these mild solutions, let us start with Leray's weak solutions which correspond to the Banach space $E = L^2(\mathbb{R}^3)$.

The following two sections do not contain original results. Their goal is to pave the way to Federbush's program. This program is a wavelet-based approach to Navier-Stokes equations.

3 Weak solutions of Navier Stokes equations

The definition of a weak solution relies on a conservation law which is satisfied by the solutions of Navier-Stokes equations : some energy is non increasing.

We start with an elementary remark

Lemma 1. *Let $a_1 = a_1(x)$, a_2 and a_3 be three real-valued functions belonging to $L^1_{\text{loc}}(\mathbb{R}^3)$. Let us assume*

$$\partial_1 a_1 + \partial_2 a_2 + \partial_3 a_3 = 0 \tag{3.1}$$

in the distributional sense and consider the differential operator

$$X = a_1\partial_1 + a_2\partial_2 + a_3\partial_3 \,. \tag{3.2}$$

Then

$$X^* = -X \tag{3.3}$$

where X^* is defined by $< X^* f, g > = < f, Xg >$ and $< f, g > = \int_{\mathbb{R}^3} f(x)\,\overline{g(x)}\,dx$.

This follows from a trivial integration by parts.

Let us now assume that $u(x,t)$ is a classical solution of Navier-Stokes equations. We furthermore assume that $u(x,t)$ satisfies together with its derivatives in x or t an estimate of the form $0(|x|^{-2})$ as $|x| \to +\infty$. A similar assumption is made on the pressure $p(x,t)$ and its gradient ∇p. We then have

Lemma 2. *For any such classical solution of the Navier-Stokes equations and for every $t \geq 0$, we have*

$$\|u(\cdot,t)\|_2^2 + 2\int_0^t \|\nabla u(\cdot,s)\|_2^2\,ds = \|u(\cdot,0)\|_2^2 \,. \tag{3.4}$$

Here and in what follows we are assuming that the viscosity ν equals 1. We will return to this issue when the scaling properties of the Navier-Stokes equations will be described. It will be shown that an obvious rescaling permits to reduce the general case to this special situation. Let us prove lemma 2.

One writes $X = u_1\partial_1 + u_2\partial_2 + u_3\partial_3$ and the Navier-Stokes equations read

$$\begin{cases} \dfrac{\partial u}{\partial t} = \Delta u - Xu - \nabla p \\ \operatorname{div} u = 0 \\ u(x,0) = u_0(x)\,. \end{cases} \tag{3.5}$$

The next lemma to be used is the following remark.

Lemma 3. *If $u = u(x) = (u_1(x), u_2(x), u_3(x))$ with $u_j \in L^2(\mathbb{R}^3)$ and $\operatorname{div} u = 0$ in the distributional sense, then for any real valued function $p(x)$ such that $\nabla p \in L^2(\mathbb{R}^3)$, we have*

$$< \nabla p, u >= 0\,. \tag{3.6}$$

It suffices to approximate p by a sequence p_j of functions in the Schwartz class such that

$$\|\nabla p - \nabla p_j\|_2 \to 0 \qquad (j \to +\infty)\,.$$

Proving $< \nabla p_j, u >= 0$ is a trivial check since $\int \partial_1\, p_j\, u_1\, dx = -\int p_j\, \partial_1\, u_1\, dx$ and so on.

Returning to lemma 2, we compute the three integrals

$$\int_{\mathbb{R}^3} u \cdot \Delta u\, dx = I_1(t) \quad , \quad \int_{\mathbb{R}^3} Xu \cdot u\, dx = I_2(t)$$

and

$$\int_{\mathbb{R}^3} \nabla p \cdot u\, dx = I_3(t)\,.$$

We then obtain $I_1(t) = -\int_{\mathbb{R}^3} |\nabla u|^2\, dx$. The integrations by parts are justified by the decay properties we imposed on u and its derivatives. Lemma 1 yields $I_2(t) = 0$ and lemma 3 implies $I_3(t) = 0$. Moreover

$$< \frac{\partial u}{\partial t}, u >= \frac{1}{2}\frac{d}{dt}\|u\|_2^2\,.$$

These computations yield

$$\frac{1}{2}\frac{d}{dt}\|u\|_2^2 = -\int |\nabla u|^2\, dx = -\|\nabla u\|_2^2 \tag{3.7}$$

which immediately implies (3.6).

As usual in mathematical physics, this conservation law leads to a natural functional space F. The solutions to Navier-Stokes equations should belong to this functional space and the problem of existence and uniqueness should be solved inside this framework. A solution belonging to this functional space F will be named a weak solution.

From (3.9), we shall demand the following two properties to a weak solution

$$\sup_{t\geq 0} \|u(\cdot,t)\|_2 \qquad \text{is finite} \tag{3.8}$$

$$\int_0^\infty \|\nabla u(\cdot,t)\|_2^2\, dt \qquad \text{is finite} \tag{3.9}$$

and we immediately warn the reader that a third condition should be imposed since (3.10) and (3.11) do not permit to define $u(\cdot,0)$.

Let us remind the reader with the definition of the homogeneous Sobolev space $\dot{H}^1(\mathbb{R}^3)$.

Definition 4. *A function $f(x)$ belongs to $\dot{H}^1(\mathbb{R}^3)$ if and only if $f(x)$ belongs to $L^6(\mathbb{R}^3)$ and ∇f belongs to $L^2(\mathbb{R}^3)$.*
The norm of f in $\dot{H}^1(\mathbb{R}^3)$ is $\|\nabla f\|_2$.

This definition looks surprising since the L^6-norm does not enter into the definition of the $\dot{H}^1(\mathbb{R}^3)$ norm. Indeed the assumption $f \in L^6$ can be replaced by a much weaker one as

$$\int_{|x|\leq R} |f(x)|\, dx = o(R^3)\,, \qquad R \to +\infty \tag{3.10}$$

which prevents constant functions to belong to $\dot{H}^1(\mathbb{R}^3)$.

The ordinary Sobolev space $H^1(\mathbb{R}^3)$ is defined by the two conditions $f \in L^2(\mathbb{R}^3)$ and $\nabla f \in L^2(\mathbb{R}^3)$.

Using these notations, we observe that a weak solution should belong to $L^\infty([0,\infty); L^2(\mathbb{R}^3))$ and to $L^2([0,\infty); \dot{H}^1(\mathbb{R}^3))$.

As it was said before, a third condition is needed for defining weak solutions. For unveiling this third condition Sobolev spaces with negative indices are needed.

Definition 5. *The Sobolev space $H^{-1}(\mathbb{R}^3)$ is the space of tempered distributions f whose Fourier transforms $\hat{f}(\xi)$ locally belong to $L^2(\mathbb{R}^3)$ and satisfy*

$$\int_{\mathbb{R}^3} |\hat{f}(\xi)|^2 (1+|\xi|^2)^{-1}\, d\xi < \infty . \tag{3.11}$$

From this definition, it is trivially checked that $H^{-1}(\mathbb{R}^3)$ is the dual space of $H^1(\mathbb{R}^3)$. More precisely $H^1(\mathbb{R}^3)$ is a substitute for the space of testing functions. When $\mathcal{S}(\mathbb{R}^3)$ is replaced by $H^1(\mathbb{R}^3)$, then $\mathcal{S}'(\mathbb{R}^3)$ is replaced by $H^{-1}(\mathbb{R}^3)$.

We now arrive to the definition of a weak solution.

Definition 6. *If $T > 0$, a weak solution $u(x,t)$ of the Navier-Stokes equations is defined by the following three properties*

$$u(x,t) \in L^\infty([0,T]; L^2(\mathbb{R}^3)) \tag{3.12}$$

$$u(x,t) \in L^2([0,T]; H^1(\mathbb{R}^3)) \tag{3.13}$$

and

$$\frac{\partial u}{\partial t} \in L^1([0,T]; H^{-1}(\mathbb{R}^3)) . \tag{3.14}$$

Similarly the pressure should belong to $L^1([0,T]; L^2(\mathbb{R}^3))$ and the Navier-Stokes equations should read

$$\frac{\partial u}{\partial t} = \Delta u - \partial_1(u_1 u) - \partial_2(u_2 u) - \partial_3(u_3 u) - \nabla p . \tag{3.15}$$

The meaning of (3.15) needs to be clarified as well as the meaning of the initial condition $u(x,0) = u_0(x)$.

We now want to motivate the definition of such weak solutions by proving some *a priori* estimates.

Lemma 4. *If $u(x,t)$ is a classical solution of the Navier-Stokes equations, we have*

$$\sup_{0\le t\le T} \|u(\cdot,t)\|_2 \le \|u_0\|_2 \tag{3.16}$$

$$\int_0^T \|u(\cdot,t)\|^2_{H^1(\mathbb{R}^3)}\,dt \le \left(T+\frac{1}{2}\right)\|u_0\|_2^2 \tag{3.17}$$

$$\int_0^T \|p(\cdot,t)\|_2\,dt \le C(T)\,\|u_0\|_2 \tag{3.18}$$

and finally

$$\int_0^T \left\|\frac{\partial u}{\partial t}(\cdot,t)\right\|_{H^{-1}(\mathbb{R}^3)} dt \le C'(T)\,\|u_0\|_2\,. \tag{3.19}$$

Let us first show that the pressure can be easily computed from the velocity field $u(x,t)$ whenever both p and u are smooth and tend to 0 in a suitable way.

In order to make life easier, we introduce the Riesz transformations R_1, R_2, R_3. They are defined by

$$R_j = -i\,\partial_j(-\Delta)^{-1/2} \qquad \text{where} \quad \partial_j = \frac{\partial}{\partial x_j}\,. \tag{3.20}$$

If $\hat{f}(\xi)$ denotes the Fourier transform of f, we have

$$(R_j f)\hat{}(\xi) = \frac{\xi_j}{|\xi|}\,\hat{f}(\xi)\,. \tag{3.21}$$

Finally R_j is a Calderón-Zygmund operator and we have

$$R_j f(x) = c \quad p.v. \int \frac{x_j - y_j}{|x-y|^4}\, f(y)\,dy \tag{3.22}$$

where c is a constant.

The Riesz transformations continuously map $L^p(\mathbb{R}^3)$ into itself when $1 < p < \infty$ and, when properly defined, continuously map $L^\infty(\mathbb{R}^3)$ into $BMO(\mathbb{R}^3)$ [32].

Lemma 5. *If (u,p) is a classical solution of the Navier-Stokes equations, we have*

$$p = R_j R_k(u_j u_k)\,. \tag{3.23}$$

Here and in what follows, $\alpha_j\beta_j$ means $\sum_1^3 \alpha_j\beta_j$. For proving this lemma, we compute the divergence of all terms in (2.1). Since $\operatorname{div} u = 0$, we also have $\operatorname{div} \frac{\partial u}{\partial t} = 0$ and $\operatorname{div} \Delta u = 0$. We then obtain

$$\Delta p + \partial_j \partial_k (u_j u_k) = 0. \tag{3.24}$$

Since we assumed that both p and u_j tend to zero at infinity, this implies (3.23).

Let us stress that R_j is not a local operator. This means that the calculation of the pressure at a given point x needs a global knowledge of the velocity field.

If $u(x,t)$ is a classical solution of Navier-Stokes equations (3.16) follows from (3.4). Similarly $\|u(\cdot,t)\|^2_{H^1(\mathbb{R}^3)} = \|u(\cdot,t)\|_2^2 + \|\nabla u(\cdot,t)\|_2^2$ and (3.17) also follows from (3.4).

The pressure estimate is slightly deeper. Indeed one uses $\|f\|_6 \le C\|\nabla f\|_2$ for $f \in \dot{H}^1(\mathbb{R}^3)$. Therefore $\int_0^T \|u(\cdot,t)\|_q^2\, dt \le C\|u_0\|_2^2$ for $2 \le q \le 6$. Hlder's inequality yields

$$\int_0^T \left\|u_j(\cdot,t)\, u_k(\cdot,t)\right\|_r dt \le C\|u_0\|_2^2 \tag{3.25}$$

for $1 \le r \le 3$. Here $u = (u_1, u_2, u_3)$.

Finally the pressure can be computed by (3.23). Since the Riesz transformations are bounded on $L^r(\mathbb{R}^3)$ for $1 < r < \infty$, we obtain

$$\int_0^T \|p(\cdot,t)\|_r\, dt \le C\|u_0\|_2^2. \tag{3.26}$$

We finally turn to the proof of (3.19). Returning to the Navier-Stokes equations, we scan Δu, ∇p and $u_1\partial_1 u + u_2\partial_2 u + u_3\partial_3 u = \partial_1(u_1 u) + \partial_1(u_2 u) + \partial_3(u_3 u)$.

Since $\Delta : H^1(\mathbb{R}^3) \to H^{-1}(\mathbb{R}^3)$ is a contraction, we have

$$\int_0^T \|\Delta u\|^2_{H^{-1}(\mathbb{R}^3)}\, dt \le \int_0^T \|u\|^2_{H^1}\, dt \le \frac{\|u_0\|_2^2}{2}.$$

Concerning the pressure, (3.26) is used with $r = 2$ and yields the required estimate.

Finally $\partial_1(u_1 u) + \partial_2(u_2 u) + \partial_3(u_3 u)$ is estimated in $L^1([0,T]; H^{-1}(\mathbb{R}^3))$ using (3.25) with $r = 2$. Lemma 4 is now proved.

We return to definition and specify the meaning of (3.17).

Both sides of (3.17) belong to $L^1([0,T]; H^{-1}(\mathbb{R}^3))$. Indeed the left-hand side does belong by (3.16). Concerning the right-hand side, it suffices to repeat the above discussion and to use (3.14) or (3.15).

Finally $u(\cdot,t)$ belongs to $C([0,T]; H^{-1}(\mathbb{R}^3))$ (indeed $u(\cdot,t)$ is absolutely continuous with respect to the time variable). Therefore $u(\cdot,0)$ can be given a meaning in the distributional sense.

Let us insist on the fact that the right-hand side of (3.17) cannot be written $u_1\partial_1 u + \cdots + u_3\partial_3 u$. Indeed one cannot multiply a function in $H^{-1}(\mathbb{R}^3)$ by a function in $L^2(\mathbb{R}^3)$. This type of problems will be answered by the paraproduct algorithm of section 14.

The existence of weak solutions will be proved in the next section in relation with the variational formulation of the Navier-Stokes equations.

4 A variational formulation of the Navier-Stokes equations

Following notations which were introduced by J.L. Lions, V will denote the closed subspace of $(H^1(\mathbb{R}^3))^3$ consisting of all vector fields $v(x) = \big(v_1(x), v_2(x), v_3(x)\big)$ such that

$$\operatorname{div} v \;=\; \partial_1 v_1 + \partial_2 v_2 + \partial_3 v_3 \;=\; 0\,. \tag{4.1}$$

The variational formulation of the Navier-Stokes equations consists in making (3.17) more explicit by writing all scalar products with all testing functions. These testing functions are functions of the x variable since we always single out the dependance in the time variable and consider $u(x,t)$ as a vector valued function of t (and not a vector valued distribution with respect to t).

But instead of using all testing functions in $\mathcal{S}(\mathbb{R}^3)$ we want instead to use testing functions in $H^1(\mathbb{R}^3)$. In other words we replace the duality between $\mathcal{S}(\mathbb{R}^3)$ and $\mathcal{S}'(\mathbb{R}^3)$ by the duality between $H^1(\mathbb{R}^3)$ and $H^{-1}(\mathbb{R}^3)$.

Moreover we indeed use $v \in V$ as testing functions. The variational formulation of the Navier-Stokes equations relies on the following lemma.

Lemma 6. *If $p(x,t) \in L^1([0,T]; L^2(\mathbb{R}^3))$ and $v \in V$, we then have*

$$< \nabla p, v > = 0 \qquad on \quad [0,T]\,. \tag{4.2}$$

Let us first observe that $< \nabla p, v > \in L^1[0,T]$. Indeed ∇p belongs to $L^1([0,T]; H^{-1}(\mathbb{R}^3))$ while v belongs to $H^1(\mathbb{R}^3)$. Since H^{-1} is the dual space of $H^1(\mathbb{R}^3)$, we can conclude.

For proving (4.2) it then suffices to use an approximation $p_j(x,t)$ to $p(x,t)$. We can assume p_j, $p_j(x,t)$ is a continuous function of the time variable with values in $\mathcal{S}(\mathbb{R}^3)$ together with

$$\int_0^T \|p(\cdot,t) - p_j(\cdot,t)\|_2 \, dt \to 0 \qquad (j \to +\infty). \tag{4.3}$$

Then

$$< \nabla p_j, v > = - < p_j, \operatorname{div} v > = 0 .$$

It suffices to pass to the limit as j tends to infinity to obtain (4.2).

Definition 7. *Let $u = u(x,t)$, $x \in \mathbb{R}^3$, $0 \le t \le T$, be a weak solution of Navier-Stokes equations. Then the variational formulation of* (3.15) *reads*

$$\begin{cases} \dfrac{d}{dt} < u, v > = < u, \Delta v > + < u, Xv > \\ \operatorname{div} u = 0 \\ u(x,0) = u_0(x) . \end{cases} \tag{4.4}$$

Here and in what follows $< u, v > = \sum_1^3 \int u_j(x) \overline{v_j(x)} \, dx$ and $X = u_1 \frac{\partial}{\partial x_1} + u_2 \frac{\partial}{\partial x_2} + u_3 \frac{\partial}{\partial x_3}$.

Let us be more specific. All terms in the first equation belong to $L^1[0,T]$.

Indeed $\frac{\partial u}{\partial t}$ belongs to $L^1([0,T]; H^{-1}(\mathbb{R}^3))$ while v belongs to $H^1(\mathbb{R}^3)$. Then $< \frac{\partial u}{\partial t}, v > \in L^1[0,T]$ and a simple limiting argument yields

$$< \frac{\partial u}{\partial t}, v > = \frac{d}{dt} < u, v > . \tag{4.5}$$

Concerning $< u, \Delta v >$, we observe that $u \in L^2([0,T]; H^1(\mathbb{R}^3))$ while $\Delta v \in H^{-1}(\mathbb{R}^3)$. Therefore $< u, \Delta v > \in L^2[0,T]$.

Finally $u \in L^2([0,T]; L^4(\mathbb{R}^3))$ which implies $Xv \in L^2([0,T]; L^{4/3}(\mathbb{R}^3))$. Since $u \in L^2([0,T]; L^4)$ we obtain $< u, Xv > \in L^1([0,T])$.

Let us conversely assume (3.12), (3.13), (3.14) and (4.4). We would like to obtain (3.15). Indeed $R(x,t) = \frac{\partial u}{\partial t} - \Delta u + Xu$ satisfies $< R, v > = 0$ for each $v \in V$. This implies $R = \nabla p$ for some scalar valued function p and Navier-Stokes equations are recovered.

For us be more explicit and consider a distribution $S = (S_1, S_2, S_3)$, $S_j \in \mathcal{S}'(\mathbb{R}^3)$, such that for any $v = (v_1, v_2, v_3)$ with $v_j \in \mathcal{S}(\mathbb{R}^3)$ and $\operatorname{div} v = 0$ we have $< S, v >= 0$. Then $S = \nabla p$. Indeed we check $< S, v >= 0$ on $v_1 = -\partial_2 f$, $v_2 = \partial_1 f$, $v_3 = 0$, $f \in \mathcal{S}(\mathbb{R}^3)$ and obtain $\partial_1 S_2 - \partial_2 S_1 = 0$ in the distributional sense. Similarly $\partial_2 S_3 - \partial_3 S_2 = 0$ and $\partial_3 S_1 - \partial_1 S_3 = 0$. This implies $S = \nabla p$ as announced.

Once the variational formulation of the Navier-Stokes equations is written, one would like to move one step further.

Instead of writing (4.32) for every $v \in V$, one is using a sequence $v_0, v_1, \ldots, v_m, \ldots$ of vectors of V with the following two properties

(4.6) $\quad v_0, v_1, \ldots, v_m, \ldots$ are linearly independant

(4.7) $\quad$ the linear span of $v_0, v_1, \ldots, v_m, \ldots$ is dense in V.

We then denote by V_m the linear span of $v_0, v_1, \ldots, v_{m-1}$. We then have $V_m \subset V_{m+1}$ and $\bigcup_{m \geq 0} V_m$ is dense in V.

We now arrive to the definition of a Galerkin scheme.

We want to approximate the weak solution $u(x,t)$ by a function $u_N(x,t)$ which is no longer a solution but belongs to $C^\infty([0,T], V_N)$.

Moreover this approximation $u_N(x,t)$ is not the orthogonal projection of $u(x,t)$ onto V_N. This implies that the convergence of u_N to u will be a delicate issue.

Instead we define $u_{0,N}(x) \in V_N$ as the orthogonal projection of u_0 onto V_N. This refers to the standard inner product in $L^2(\mathbb{R}^3)$.

We now define $u_N(x,t)$ by

$$\frac{d}{dt} < u_N, v_m > = < u_N, \Delta v_m > + < u_N, (u_N \cdot \nabla) v_m > \tag{4.8}$$

where $0 \leq m \leq N-1$ and

$$u_N(x, 0) = u_{0,N}(x) . \tag{4.9}$$

Let us simplify the discussion by assuming the following. If H_0 denotes the closed subspace of $(L^2(\mathbb{R}^3))^3$ defined by $\partial_1 f_1(x) + \partial_2 f_2(x) + \partial_3 f_3(x) = 0$, then v_m, $m \geq 0$, is a Hilbert space basis of H_0. We then write

$$u_N(x,t) = \alpha_{0,N}(t)\, v_0(x) + \cdots + \alpha_{N-1,N}(t)\, v_{N-1}(x) \tag{4.10}$$

and (4.8) reads

$$\frac{d}{dt}\alpha_N(t) = A[\alpha_N] + B[\alpha_N, \alpha_N] \tag{4.11}$$

where $\alpha_N(t) = \big(\alpha_{0,N}(t), \dots, \alpha_{N-1,N}(t)\big)$, A being an ordinary $N \times N$ matrix and $B[X, X]$ and ordinary quadratic form. Obviously A and B depend on N.

This ordinary differential equation should be completed with $\alpha_N(0) = \gamma_N$ where $\gamma_N = \big(\gamma_0, \dots, \gamma_{N-1}\big)$ when $u_0(x) = \sum_0^\infty \gamma_m v_m(x)$.

Since (4.11) is non linear, the only issue concerns a possible blow up in finite time.

However the proof of the energy estimate can be rewritten inside the approximation space V_N and yields

$$\|u_N(\cdot, t)\|_2 \le \|u_{0,N}\|_2 \le \|u_0\|_2 \,. \tag{4.12}$$

It implies that (4.8) has a unique global solution and $u_N(x,t)$ belongs to $C^\infty([0,T]; V_N)$ for all positive T's.

This being achieved, one would like to build a true solution from this sequence $u_N(x,t)$, $N \ge 1$. Since u_N uniformly satisfies the estimates listed in lemma 4, one is tempted to use a compactness argument. From $u_N(x,t)$ one is extracting a subsequence $u_{N_j}(x,t)$ which weakly converges to $u(x,t)$. For proving that $u(x,t)$ is a solution, one should intertwin weak limits and non linear mappings. We all know that it is impossible and our naive approach fails. We then follow J.L. Lions and construct a subsequence $u_{N_j}(x,t)$ such that

$$\lim_{j\to+\infty} \int_0^T \int_{|x|\le R} \left|u_{N_j}(x,t) - u(x,t)\right|^2 dx\, dt = 0 \tag{4.13}$$

for every positive R.

For proving this strong convergence, a new estimate is needed.

J.L. Lions proves the existence of a positive exponent γ and of a constant $C(\gamma)$ such that, for any $u_0(x)$ in $L^2(\mathbb{R}^3)$ and $N \ge 1$, one has

$$\int_{-\infty}^{\infty} |\tau|^{2\gamma}\, \|\hat{u}_N(\cdot, \tau)\|_2^2\, d\tau \le C(\gamma)\, \|u_0\|_2^2 \,. \tag{4.14}$$

Here $\hat{u}_N(x,\tau) = \int_{-\infty}^{\infty} e^{-it\tau}\, u_N(x,t)\, dt$ and the L^2 norm is computed with respect to the x variable. This new estimate, once combined with

$$\sup_{t\ge 0} \|u_N(\cdot,t)\|_2 \le C_0 \quad \text{and} \quad \int_0^T \|u_N(\cdot,t)\|_{H^1}^2\, dt \le C_1$$

is used for showing that the weak convergence in $L^2([0,T]; H^1(\mathbb{R}^3))$ of u_{N_j} implies (4.41).

The reader is referred to [65] and J. Leray's theorem will be stated.

Theorem 4.1. *Let us assume that* $u_0(x) \in L^2(\mathbb{R}^3)$ *and* $\operatorname{div} u_0(x) = 0$. *Then for every* $T > 0$, *there exists a weak solution* $u(x,t)$, $0 \le t \le T$, *of the Navier-Stokes equations such that* $u(x,0) = u_0(x)$.

The uniqueness of such a weak solution is still an open problem.

5 The affine group action

The Navier-Stokes equations are invariant under a group action. More precisely if $(u(x,t), p(x,t))$ is a solution defined on $\mathbb{R}^3 \times (0,\infty)$, the same is true for (u_λ, p_λ) where $0 < \lambda < \infty$ and

$$(5.1) \qquad u_\lambda(x,t) = \lambda u(\lambda x, \lambda^2 t) \quad , \quad p_\lambda(x,t) = \lambda^2 p(\lambda x, \lambda^2 t)\,.$$

If instead one would consider $u(\lambda x, \lambda t)$ and $p(\lambda x, \lambda t)$, there are a solution of a distinct Navier-Stokes equation where ν is being replaced by $\lambda\nu$. We can fix $\lambda > 0$ such that $\lambda\nu = 1$ and assume that the viscosity is 1.

Returning to the invariance of the Navier-Stokes equation, if $(u(x,t), p(x,t))$ is a solution, the same is true for $\big(u(x-x_0,t), p(x-x_0,t)\big)$ for $x_0 \in \mathbb{R}^3$.

Finally for any $\tau \ge 0$, $\big(u(x,t+\tau), p(x,t+\tau)\big)$ is still a solution.

This affine group action should be incorporated into a Galerkin scheme. Guy Battle and Paul Federbush achieved this program and their results will now be described.

Let us denote by H_0 the subspace of $H = (L^2(\mathbb{R}^3))^3$ which is defined by $\partial_1 u_1 + \partial_2 u_2 + \partial_3 u_3 = 0$. Since this condition can be rewritten as $R_1 u_1 + R_2 u_2 + R_3 u_3 = 0$ where R_j are the Riesz transformations, H_0 is a closed subspace of H. G. Battle and P. Federbush succeeded in constructing a basis for H_0 of the form

$$(5.2) \qquad 2^{3j/2}\,\psi(2^j x - k)\,, \qquad \psi \in A\,,\ j \in \mathbf{Z}\,,\ k \in \mathbf{Z}^3$$

where A is a finite collection of 14 mother wavelets belonging to the Schwartz class.

Before moving further, let us sketch the construction of Federbush's basis. We closely follow the modification proposed by P.G. Lemarié-Rieusset in [50] of Federbush's original construction.

6 A divergence-free wavelet basis

The starting point is the orthonormal wavelet basis of $L^2(\mathbb{R})$ which is presented in [55]. We begin with the so-called scaling function $\varphi(t)$ which is a smooth version of the indicator function of $[0,1]$. Here $\varphi(t)$ belongs to the Schwartz class, its Fourier transform $\hat{\varphi}(\xi)$ satisfies $\hat{\varphi}(\xi) = 1$ on $[-2\pi/3\,,\,2\pi/3]$, $\hat{\varphi}(\xi) = 0$ if $|\xi| \geq 4\pi/3$, $\hat{\varphi}(\xi) \in [0,1]$, $\hat{\varphi}(-\xi) = \hat{\varphi}(\xi)$ and finally

$$|\hat{\varphi}(\pi+s)|^2 + |\hat{\varphi}(\pi-s)|^2 = 1 \quad , \quad |s| \leq \pi/3\,. \tag{6.1}$$

These assumptions imply $\sum_{-\infty}^{\infty} |\hat{\varphi}(2k\pi+\xi)|^2 = 1$ everywhere. Therefore $\varphi(t-k)$, $k \in \mathbf{Z}$, is an orthonormal sequence in $L^2(\mathbb{R})$. We denote by V_0 the closed linear span of this sequence $\varphi(t-k)$, $k \in \mathbf{Z}$. Next V_j is defined by

$$f(t) \in V_0 \iff 2^{j/2} f(2^j t) \in V_j \tag{6.2}$$

where $f \in L^2(\mathbb{R})$ and $j \in \mathbf{Z}$.

A crucial observation is the inclusion $V_j \subset V_{j+1}$ which relies on the specific properties of φ.

The next step consists in studying the orthogonal complement W_j of V_j inside V_{j+1}. Property (6.2) implies

$$f(t) \in W_0 \iff 2^{j/2} f(2^j t) \in W_j\,. \tag{6.3}$$

There exists a function $\psi(t)$ in the Schwartz class such that $\psi(t-k)$, $k \in \mathbf{Z}$, is an orthonormal basis of W_0.

The Fourier transform $\hat{\psi}(\xi)$ of ψ is supported by $\frac{2\pi}{3} \leq |\xi| \leq \frac{8\pi}{3}$ and

$$2^{j/2}\,\psi(2^j t-k)\,, \qquad j \in \mathbf{Z}\,,\ k \in \mathbf{Z} \tag{6.4}$$

is an orthonormal basis for $L^2(\mathbb{R})$. This function $\psi(t)$ is the mother wavelet.

Both the scaling function $\varphi(t)$ and the mother wavelet $\psi(t)$ are needed to construct an orthonormal wavelet basis for $L^2(\mathbb{R}^3)$.

Indeed one writes $\varphi_0(t) = \varphi(t)$, $\varphi_1(t) = \psi(t)$ and the three-dimensional scaling function is

$$\varphi(x) = \varphi_0(x_1)\,\varphi_0(x_2)\,\varphi_0(x_3) \tag{6.5}$$

while the 7 three-dimensional wavelets are given by

$$\psi_\varepsilon(x) = \varphi_{\varepsilon_1}(x_1)\,\varphi_{\varepsilon_2}(x_2)\,\varphi_{\varepsilon_3}(x_3) \quad , \quad \varepsilon = (\varepsilon_1, \varepsilon_2, \varepsilon_3) \tag{6.6}$$

and $\varepsilon \in \{0,1\}^3$, $\varepsilon \neq (0,0,0)$.

With these notations, we obtain the three-dimensional wavelet basis as

$$\begin{aligned} 2^{3j/2}\,\psi_\varepsilon(2^j x - k)\,, \qquad & j \in \mathbf{Z}\,,\ k \in \mathbf{Z}^3\,, \\ \varepsilon \in \{0,1\}^3 \quad , \quad & \varepsilon \neq (0,0,0)\,. \end{aligned} \tag{6.7}$$

We now denote by H_0 the closed subspace of $H = (L^2(\mathbb{R}^3))^3$ defined by $R_1u_1 + R_2u_2 + R_3u_3 = 0$ where R_1, R_2, R_3 are the three Riesz transformations. Each component u_1, u_2 or u_3 is then expanded into a wavelet expansion using the orthonormal basis given by (6.51).

Roughly speaking, the full equation $R_1u_1 + R_2u_2 + R_3u_3 = 0$ can be decoupled into

$$R_1\, u_1^{(j,\varepsilon)} + R_2\, u_2^{(j,\varepsilon)} + R_3\, u_3^{(j,\varepsilon)} = 0 \tag{6.8}$$

where $u_1^{(j,\varepsilon)}$ belongs to the closed linear span of $2^{3j/2}\,\psi_\varepsilon(2^j x - k)$, $k \in \mathbf{Z}^3$.

Then if $\varepsilon_1 = 1$, (6.52) permits to compute $u_1^{(j,\varepsilon)}$ as a linear combination of $u_2^{(j,\varepsilon)}$ and $u_3^{(j,\varepsilon)}$. Since one of the three indices $\varepsilon_1, \varepsilon_2$ or ε_3 is 1, (6.52) can always be solved.

This construction yields a basis for H_0 which is no longer an orthogonal one. Let us clarify this point.

Definition 8. *A Riesz basis $(e_\lambda)_{\lambda\in\Lambda}$ in a Hilbert space H is a collection of vectors of H for which there exists an orthonormal basis $(f_\lambda)_{\lambda\in\Lambda}$ of H and an isomorphism $U : H \to H$ such that*

$$U(f_\lambda) = e_\lambda\,. \tag{6.9}$$

It implies that each vector $x \in H$ can uniquely be written as

$$x = \sum_{\lambda\in\Lambda} \alpha_\lambda\, e_\lambda \tag{6.10}$$

where

$$C_1\|x\| \le \left(\sum |\alpha_\lambda|^2\right)^{1/2} \le C_2\|x\| \tag{6.11}$$

for two constants $C_2 \geq C_1 > 0$.

If $C_2 = C_1 = 1$, $(e_\lambda)_{\lambda \in \Lambda}$ is an orthonormal basis. In general, one defines the dual basis $\tilde{e}_\lambda$ by

$$(U^*)^{-1}(f_\lambda) = \tilde{e}_\lambda \,. \tag{6.12}$$

This dual basis is still a Riesz basis and one has

$$\alpha_\lambda = <x, \tilde{e}_\lambda> \,. \tag{6.13}$$

With this definition in mind, the Lemarié-Rieusset's version of the Battle-Federbush theorem reads

Theorem 6.1. *There exists a finite collection A of 14 divergence free vector fields $\vec{\psi}(x) = \big(\psi_1(x), \psi_2(x), \psi_3(x)\big)$ with the following properties*

$$\psi_j \in \mathcal{S}(\mathrm{I\!R}^3)\,, \qquad 1 \leq j \leq 3 \tag{6.14}$$

$$\hat{\psi}_j(\xi) = 0 \quad \textit{if} \quad |\xi|_\infty \leq \frac{2\pi}{3} \quad \textit{or} \quad |\xi|_\infty \geq \frac{8\pi}{3} \tag{6.15}$$

where $|\xi|_\infty = \sup\big(|\xi_1|, |\xi_2|, |\xi_3|\big)$

$$2^{3j/2}\,\psi(2^j x - k)\,, \qquad \psi \in A\,,\ j \in \mathbf{Z}\,,\ k \in \mathbf{Z}^3\,, \tag{6.16}$$

is a Riesz basis of H_0

(6.17) *the dual basis is given by*

$$2^{3j/2}\,\tilde{\psi}(2^j x - k)\,, \qquad \tilde{\psi} \in \tilde{A}\,,\ j \in \mathbf{Z}\,,\ k \in \mathbf{Z}^3$$

where $\tilde{A}$ is a second collection of 14 wavelets satisfying (6.14) *and* (6.15).

This remarkable theorem will be proved in the appendix and we are now ready for describing P. Federbush's program.

7 Federbush's program

Federbush's program is quite natural and extremely appealing. It consists in using a Galerkin scheme for solving Navier-Stokes equations and in incorporating the affine group action into this Galerkin scheme. We then respect the invariance of the Navier-Stokes equations with respect to this affine group action. This approach is consistent with our scientific knowledge about turbulence. We will return to this point.

A Galerkin scheme where affine group action is incorporated is the Battle-Federbush basis. This basis reads $2^{3j/2}\,\vec{\psi}(2^jx-k)$, $j \in \mathbf{Z}$, $k \in \mathbf{Z}^3$, $\vec{\psi} \in A$ where A is a collection of 14 divergence free mother wavelets $\vec{\psi} = (\psi_1, \psi_2, \psi_3)$ for which ψ_1, ψ_2 and ψ_3 belong to the Schwartz class. Such an approach is aimed to decouple the Navier-Stokes equation into a sequence of equations of the general form

$$\frac{\partial}{\partial t} u_j = L_j(u_j) + \sum_{j'}\sum_{j''} B_{j,j',j''}(u_{j'}, u_{j''})\,. \tag{7.1}$$

Here $u_j(x,t) = \sum_k \alpha(j,k,t)\, 2^{3j/2}\,\vec{\psi}(2^jx-k)$ (the sum over A will be omitted for keeping the notations as short as possible).

Since the Fourier transform of $2^{3j/2}\,\vec{\psi}(2^jx-k)$ is supported by the dyadic annulus

$$\frac{2\pi}{3}\,2^j \le |\xi|_\infty \le \frac{8\pi}{3}\,2^j \tag{7.2}$$

the closed linear span of $2^{3j/2}\,\vec{\psi}(2^jx-k)$, $\vec{\psi} \in A$, $k \in \mathbf{Z}^3$, will be named a frequency channel and denoted by $\vec{W}_j$, $j \in \mathbf{Z}$.

Returning to Federbush's program, the goal consists in decoupling the Navier-Stokes equations into a sequence of equations. This decoupling only applies to the linear term (Δu) while the non-linear terms $\big(\partial_1(u_1u) + \cdots + \partial_3(u_3u)\big)$ are coupled by non-linear interactions. These non-linear interactions are described by the bilinear operators $B_{j,j',j''}(u_{j'}, u_{j''})$.

The success of this approach depends on the following property which is expected from $B_{j,j',j''}$. These interactions should become negligible when $|j'-j| + |j''-j|$ tends to infinity.

In other words, what is happening inside a frequency channel should only affect the neighboring frequency channels.

Before entering into a more precise criticism of this program, let us observe that it is backed on some previous work by P. Frick and V. Zimin.

In a superb paper entitled *Hierarchical models of turbulence* [33] they write

> "Ideas, like the ones used to create wavelet analysis were proposed by Zimin (1981) for construction of a hierarchical model of turbulence...
>
> In a paper by Zimin (1981) a special functional basis has been presented. Functions of this basis are related to a hierarchical

system of vortices of different sizes. The number of vortices in a unit volume increases with decreasing size and each function is well localized both in Fourier and physical spaces. The product of the characteristic scales of localization in Fourier and coordinate spaces satisfies the uncertainty relation...

The cascade equations, written for the quantities A_i, each define the velocity oscillations in some interval of wave numbers and describe the principal characteristics of energy redistribution processes between different scales. The cascade equations minimize the dimensionality of systems which describe the turbulent flows in wide range of wave numbers, and have a form

$$\frac{d}{dt} A_i = Y_i A_i + \sum_{j,k}\sum X_{ijk} A_j A_k \ldots \tag{7.3}$$

The hierarchical model of turbulence is based on the natural assumption that turbulence is an ensemble of vortices of progressively diminishing scales. The hierarchical basis for two-dimensional turbulence describes the ensemble of the vortices, in which any vortex of the given size consists of four vortices of half size and so on. The ensemble of vortices of the same size forms a "level". The functions of the hierarchical basis are constructed in such a way that Fourier-images of vortices of single level occupy only single octave in the wave-number space and regions of localization of different levels in the Fourier space do not overlap.

The wave-number space is divided at ring zones such that $\pi 2^N < |k| < \pi 2^{N+1} \ldots$"

This quotation from Frick and Zimin implies that these authors are using the Shannon wavelet basis. This remark is made explicit in their paper. The scaling function of the one-dimensional Shannon basis is the standard sinc function $\varphi(t) = \frac{\sin \pi t}{\pi t}$ while the corresponding mother wavelet is given by

$$\psi(t) = 2\varphi(2t) - \varphi(t)\,. \tag{7.4}$$

However these functions have a poor localization in the coordinate space. They do have an ideal localization in the frequency domain. Shannon wavelets correspond to ideal filters in signal processing and these ideal filters are unrealistic ones since their numerical support is infinite. For the same reason, Shannon wavelets cannot be used in numerical analysis.

We now want to use "modern technology" and further analyze Federbush's program.

Let us try to decouple Navier-Stokes equations. We denote by $\psi_\lambda(x)$, $\lambda \in \Lambda$, the divergence-free wavelet basis which is described in theorem 6.1. We have $\Lambda = A \times \mathbf{Z} \times \mathbf{Z}^3$ and $\psi_\lambda(x) = 2^{3j/2}\,\psi(2^j x - k)$, $\psi \in A$, $j \in \mathbf{Z}$, $k \in \mathbf{Z}^3$, $\lambda = (\psi, j, k) \in \Lambda$.

Using the variational formulation we are led to writing

$$u(x,t) = \sum_{\lambda \in \Lambda} \alpha_\lambda(t)\,\psi_\lambda(x)\,. \tag{7.5}$$

The corresponding Galerkin scheme reads

(7.6)

$$\frac{d}{dt}\alpha_\lambda(t) = \sum_{\lambda'} \omega(\lambda,\lambda')\,\alpha_{\lambda'}(t) + \sum\sum_{\lambda',\lambda''} \beta(\lambda,\lambda',\lambda'')\,\alpha_{\lambda'}(t)\,\alpha_{\lambda''}(t)$$

where

$$\begin{aligned} \omega(\lambda,\lambda') &= \big(\Delta\psi_{\lambda'}, \tilde{\psi}_\lambda\big) \\ \beta(\lambda,\lambda',\lambda'') &= b\big(\psi_{\lambda'}, \psi_{\lambda''}, \tilde{\psi}_\lambda\big) \end{aligned}$$

and

$$b(u,v,w) = \sum_1^3\sum_1^3 \int u_k(x)\big(\partial_k v_\ell\big)(x)\,\overline{w}_\ell(x)\,dx\,.$$

Concerning $\omega(\lambda,\lambda')$, we first observe that $\omega(\lambda,\lambda') = 0$ whenever $|j'-j| \geq 2$. Since the mother wavelets belong to the Schwartz class, we also obtain a rapid decay when $|j'-j| \leq 1$ and when the distance between λ' and λ tends to infinity. This distance is defined as $(2^j+2^{j'})|k2^{-j}-k'2^{-j'}|$ if $k2^{-j} \neq k'2^{-j'}$ and $2^{|j'-j|}$ if $k2^{-j} = k'2^{-j'}$, $\psi \neq \psi'$.

It should be stressed that one cannot obtain a rapid decay at infinity if Shannon's wavelets are being used. Returning to the one-dimensional case, it is not true that $\int \varphi''(x)\,\varphi(x-k)\,dx$ has a rapid decay at infinity as $|k| \to +\infty$. Indeed this integral is $0(k^{-2})$ and this estimate is sharp.

This means that the Frick-Zimin program cannot be completed the way it is stated.

Let us forget this remark and return to Federbush's version of the Frick-Zimin approach. There is unexpected bad news given by the following claim.

Claim. *There exists a vector $\ell \in \mathbf{Z}^3$ with the following property : if $j' = j''$, $k' - k'' = \ell$ and if j tends to $-\infty$, then the trilinear coefficients $\beta(\lambda, \lambda', \lambda'')$ are large.*

More precisely, these coefficients are large whenever the dyadic cube $Q(\lambda)$ contains $Q(\lambda') \cup Q(\lambda'')$. Here $Q(\lambda)$ denotes the dyadic cube defined by $2^j x - k \in [0,1)^3$ when $\psi_\lambda(x) = 2^{3j/2}\,\psi(2^j x - k)$.

This unpleasant fact contradicts big claims made by some scientists ([28], page 664).

Let us now prove our claim. We are assuming $j' = j''$ and first want to prove that the function

$$(\psi_{\lambda'} \cdot \nabla)\psi_{\lambda''} \;=\; \psi^{(1)}_{\lambda'}\partial_1\psi_{\lambda''} + \psi^{(2)}_{\lambda'}\partial_2\psi_{\lambda''} + \psi^{(3)}_{\lambda'}\partial_3\psi_{\lambda''}$$

has a non vanishing integral. Indeed

$$I(\lambda', \lambda'') \;=\; \int (\psi_{\lambda'} \cdot \nabla)\psi_{\lambda''}\,dx \;=\; 2^{j'}\gamma(k'-k'')\,.$$

Let us show that γ does not vanish identically. Otherwise one would have

$$\int (\psi \cdot \nabla)\,\psi(x-k)\,dx \;=\; 0\,, \qquad k \in \mathbf{Z}^3\,. \tag{7.7}$$

But (7.7) is equivalent to

$$\sum_{k \in \mathbf{Z}^3} \sigma(\xi - 2k\pi) \;=\; 0 \tag{7.8}$$

where

$$\sigma(\xi) \;=\; \left[\xi_1\hat{\psi}_1(\xi) + \xi_2\hat{\psi}_2(\xi) + \xi_3\hat{\psi}_3(\xi)\right]\overline{\hat{\psi}}(\xi)\,. \tag{7.9}$$

Here we need to compute $\hat{\psi}(\xi)$, $\sigma(\xi)$ and to check that (7.9) does not hold. These computations are based on the explicit calculations made in the appendix and will be omitted. Therefore the product $(\psi_{\lambda'} \cdot \nabla)\psi_{\lambda''}$ is given by

$$(\psi_{\lambda'} \cdot \nabla)\psi_{\lambda''} \;=\; 2^{4j'}\,\omega(2^{j'}x - k') \tag{7.10}$$

where $\omega \in \mathcal{S}(\mathbb{R}^3)$, $\omega = \omega_\ell$ and $\int \omega(x)\,dx \neq 0$. Then

$$\begin{aligned}\beta(\lambda, \lambda', \lambda'') \;&=\; 2^{4j'} \int \omega(2^{j'}x - k')\,\tilde{\psi}_\lambda(x)\,dx \\ &=\; 2^{j'}\,\tilde{\psi}_\lambda(k'2^{-j'}) \cdot \left(\gamma(k'-k'') + 0(2^{j-j'})\right)\end{aligned}$$

as $j - j'$ tends to $-\infty$. This proves our claim.

This discussion will be elaborated in section 14 where the paraproduct algorithm will be studied. Indeed the paraproduct algorithm consists in writing the ordinary product $f(x)\,g(x)$ between two arbitrary functions as a sum of three series. Two of them only contain oscillatory terms. The third one is the most difficult. It reads $\sum_{-\infty}^{\infty} \Delta_j(f)\,\Delta_j(g)$ and these products do not have any cancellation. They are similar to $(\psi_{\lambda'} \cdot \nabla)\psi_{\lambda''}$ when $j' = j''$.

Let us now return to P. Federbush's program.

In a paper entitled *Navier and Stokes meet the wavelet*, P. Federbush was much hoping that our understanding of the Navier-Stokes equations could benefit from wavelet bases.

Let me first stress that I fully believe that P. Federbush's claim is correct. However Federbush's paper is disappointing.

First of all, Federbush is not proposing a new numerical scheme for solving Navier-Stokes equations. Instead he is using his wavelet basis for proving new estimates. These estimates are defined by functional norms. The corresponding functional spaces which are used in Federbush's paper are the Morrey-Campanato spaces.

Definition 9. *If $s \in (0,3)$, the Banach space $M^{2,s}$ is defined by the following property. A function $f(x)$ belongs to $M^{2,s}$ if and only if f is locally square integrable and if there exists a constant C such that*

$$\sup_{x_0 \in \mathbb{R}^3} \sup_{0<R<\infty} \left(R^{-s} \int_{|x-x_0| \le R} |f(x)|^2\, dx \right)^{1/2} \le C\,. \tag{7.11}$$

If $s = 0$, this would define L^2 and if $s = 3$, (7.72) would define L^∞.

P. Federbush restricted his attention to the case $1 < s < 3$ and proved the following statement. If $u_0(x)$ belongs to $M^{2,s}$ and $\operatorname{div} u_0(x) = 0$, then there exists a positive T and a solution

$$u(x,t) \in C\big([0,T], M^{2,s}\big) \tag{7.12}$$

of the Navier-Stokes equations such that $u(x,0) = u_0(x)$.

These solutions $u(x,t) \in C([0,T], M^{2,s})$ are no longer classical solutions of Navier-Stokes equations. They are not weak solutions in Leray's sense. The solutions constructed by P. Federbush in his paper are the so-called "mild solutions" of Navier-Stokes equations. This concept will be defined in section 12.

With my student Marco Cannone, we wanted to better understand Federbush's paper and the role played by the hierarchical wavelet basis in the proof. We soon observed that Federbush's theorem is becoming a trivial statement if a Littlewood-Paley analysis will be defined in section 14 and wavelet analysis can be defined as an improved version of Littlewood-Paley analysis.

The second piece of bad news is the following. The case $s = 1$ in the definition of the Morrey-Campanato spaces $M^{2,s}$ cannot be treated by Federbush's approach. This exponent plays a crucial role since homogeneous functions of degree -1 belong to $M^{2,1}$. If the initial condition $u_0(x)$ exhibits this homogeneity, then the corresponding solution of the Navier-Stokes equation will be self-similar. This means that

$$\lambda u(\lambda x, \lambda^2 t) = u(x,t)\,, \qquad 0 < \lambda < \infty\,. \tag{7.13}$$

In other words solving Navier-Stokes equations when $u_0 \in M^{2,1}$ paves the way to the construction of self-similar solutions to the Navier-Stokes equations. In the next section, the special role played by this Morrey-Campanato space $M^{2,1}$ will be clarified. Here we want to stress that Tosio Kato found an algorithm for solving Navier-Stokes equations when $u_0 \in M^{2,1}$. This remarkable algorithm is not based on Fourier analysis or wavelet analysis. Kato's approach will be presented in section 21.

These remarks may lead to the conclusion that wavelet analysis is the worst tool for studying Navier-Stokes equations. If Federbush's paper were the test for ranking all available tools, the conclusion would be obvious. Wavelet analysis is the worst tool, Littlewood-Paley analysis and paraproduct algorithms are doing a much better job and T. Kato's algorithm is number one.

This bad news does not end the discussion. Indeed Littlewood-Paley analysis or wavelet analysis were used in the first proof of the uniqueness of mild solutions $u \in C[0,\infty)$, of Navier-Stokes equations. The existence of such solutions was proved by T. Kato but Kato's algorithm left the uniqueness problem open. Unfortunately a much simpler proof was soon obtained. This simpler proof does not use any spectral method and the usefulness of Littlewood-Paley methods cannot be proved in this uniqueness issue.

8 Banach spaces adapted to Navier-Stokes equations

Our goal is to describe an improved version of Kato's program. T. Kato treated an important example. He proved the existence of a positive constant α such that the following implication holds : if $u_0(x)$ belongs to $L^3(\mathbb{R}^3)$, if $\|u_0\|_3 < \alpha$ and $\operatorname{div} u_0(x) = 0$, then there exists a solution

$$(8.1) \qquad u(x,t) \in C\big([0,\infty); L^3(\mathbb{R}^3)\big)$$

of the Navier-Stokes equations such that

$$(8.2) \qquad u(x,0) = u_0(x)\,.$$

The uniqueness of such a solution will be proved in section 20 and a second proof can be found in section 26. Let us observe that $u_0(x)$ and $\lambda u_0(x)(\lambda x)$, $\lambda > 0$, are sharing the same L^3 norm. In other words the exponent 3 cannot be replaced by an other one in Kato's theorem. Indeed if the initial condition $u_0(x)$ does not satisfy $\|u_0\|_q < \alpha$, then $\lambda u_0(\lambda x)$ will certainly do it for some value of λ. This means that either the condition $\|u_0\|_q < \alpha$ is not needed and can be replaced by $u_0 \in L^q(\mathbb{R}^3)$ or it is needed and $q = 3$. The first option applies to the case $q = 2$ and nobody knows what is happening for other values of q.

In our attempt to generalize Kato's program, $L^3(\mathbb{R}^3)$ will be replaced by other Banach spaces satisfying some properties which are given in definition 10 and 11.

Definition 10. *A Banach space E is a functional Banach space if*

(8.3) $\mathcal{S}(\mathbb{R}^n) \subset E \subset \mathcal{S}'(\mathbb{R}^n)$

(8.4) *these two canonical embeddings are continuous ones*

(8.5) *either these two embeddings have a dense range*
or E is the dual space F^* of a functional Banach space F
for which these two embeddings have a dense range.

For example $L^p(\mathbb{R}^n)$, $1 \le p < \infty$ is fulfilling these requirements while $L^\infty(\mathbb{R}^n)$ is the dual space of $L^1(\mathbb{R}^n)$. It is clear that $\mathcal{S}(\mathbb{R}^n) \subset L^\infty(\mathbb{R}^n)$ but $\mathcal{S}(\mathbb{R}^n)$ is not dense inside $L^\infty(\mathbb{R}^n)$.

Definition 10 paves the way to the following one.

Definition 11. *A functional Banach space E is adapted to the Navier-Stokes equations if*

(8.6) $$\|f(x)\|_E = \|\lambda f(\lambda x)\|_E\,, \qquad 0<\lambda<\infty$$

(8.7) $$\|f(x-x_0)\|_E = \|f(x)\|_E$$

(8.8) *there is a constant C such that*

$$\left(\int_{|x|\le 1} |f(x)|^2\,dx\right)^{1/2} \le C\|f\|_E\,.$$

If E is adapted to the Navier-Stokes equations, our program consists in generalizing Kato's theorem. More precisely we would like to prove the following conjecture :

Conjecture. *Let us assume that a functional Banach space E is adapted to the Navier-Stokes equations. Then there exists a positive number η such that*

(8.9) $$\|u_0\|_E < \eta \quad \textit{and} \quad \operatorname{div} u_0(x) = 0$$

implies that a solution to the Navier-Stokes equations $u(x,t)$ exists with the following properties

(8.10) $$u(x,t) \in C([0,\infty);E)$$

(8.11) $$u(x,0) = u_0(x)\,.$$

Some results in this direction will be given in section 19 and 22.

When $\mathcal{S}(\mathbb{R}^n)$ is dense in E, $C([0,\infty);E)$ will denote the Banach space consisting of all continuous and bounded functions $f:[0,\infty)\to E$. This continuity refers to the norm topology on E. The norm of f in $C([0,\infty);E)$ is defined as

(8.12) $$\sup_{t\ge 0} \|f(\cdot,t)\|_E\,.$$

When $\mathcal{S}(\mathbb{R}^n)$ is not dense in E but instead E is the dual space F^* of a functional Banach space F in which $\mathcal{S}(\mathbb{R}^n)$ is dense, it would be natural to define $C([0,\infty);E)$ as being the Banach space of continuous and bounded functions from $[0,\infty)$ into E when E is given the $\sigma(E,F)$ topology.

However we will not adopt this definition. Indeed some non-linear operations will be performed on such functions f and the $\sigma(E,F)$ topology is not consistent with such nonlinearities.

The definition of $u \in C([0,\infty);E)$ is motivated by the simple example of the heat equation

$$(8.13) \qquad \frac{\partial u}{\partial t} = \Delta u \quad , \quad u = u(x,t) \quad , \quad u(x,0) = u_0(x)$$

where $u_0 \in E$ and $\sup\left\{\|u(\cdot,t)\|_E \,;\, t \geq 0\right\}$ is finite. Then we have

$$(8.14) \qquad u(x,t) \;=\; S(t)u_0 \quad , \quad S(t) = \exp(t\Delta) \quad , \quad t \geq 0\,.$$

At $t_0 = 0$, we have $\lim_{t\downarrow 0} u(x,t) = u_0(x)$ when E is given its $\sigma(E,F)$ topology and this result is optimal when E is not a separable Banach space. However when $t_0 > 0$, $\lim_{t\to t_0} \|u(x,t) - u(x,t_0)\|_E = 0$. This leads to the following definition :

Definition 12. *When the functional Banach space E is not separable but instead is the dual F^* of a separable Banach space F, we will write*

$$(8.15) \qquad u(x,t) \;\in\; C([0,\infty);E)$$

if and only if

$$(8.16) \qquad \lim_{t\downarrow 0} u(x,t) \;=\; u(x,0) \quad \textit{when } E \textit{ is equipped}$$

with its $\sigma(E,F)$ topology and

$$(8.17) \qquad \lim_{t\to t_0} \|u(x,t) - u(x,t_0)\|_E \;=\; 0 \quad \textit{for} \quad t_0 > 0\,.$$

This definition will receive an *a posteriori* vindication in section 23. Indeed for a large family of adapted Banach spaces E, (8.91) will be improved into

$$(8.18) \qquad \|u(x,t') - u(x,t)\|_E \;\leq\; C\left|\frac{t'}{t} - 1\right|^{1/2}$$

for $0 < t < t' < 2t$ with $C = C(u_0)$.

Let us end this section with a simple but striking fact. There exist two adapted functional Banach spaces E_0 and E_1 such that any adapted functional space necessarily satisfies

$$(8.19) \qquad E_0 \subset E \subset E_1\,.$$

In other words, among adapted functional Banach spaces, there exists a maximal space E_1 and a minimal space E_0.

Let us stress that the other adapted Banach spaces are not ordered by inclusion. Simple examples will be given in section 24.

Let us first concentrate on the maximal space E_1. Let us consider the Morrey-Campanato space $M^2 = M^{2,1}$. We then have

Lemma 7. *The Morrey-Campanato space M^2 is adapted to the Navier-Stokes equations. Conversely if E is adapated to the Navier-Stokes equations, then E is continuously embedded into M^2.*

The second assertion is almost obvious. If $f \in E$ and $\|f\|_E \le 1$ then $f_{\lambda,x_0}(x) = \lambda f(\lambda(x - x_0))$ also belongs to the unit ball of E. We then apply (8.8) to f_{λ,x_0} and immediately obtain $f \in M^2$.

The proof of the first assertion is straightforward. However the Morrey-Campanato space M^2 is not separable : $\mathcal{S}(\mathbb{R}^3)$ is not dense in M^2.

Instead M^2 is the dual space of a separable Banach space F. This Banach space F admits an atomic decomposition and we first descrribe the corresponding atoms. These atoms $a_B(x)$ are labelled by balls $B \subset \mathbb{R}^3$. More precisely B is an arbitrary ball in $\mathbb{R}^3$ and $a_B(x)$ should satisfy the following two properties

$$\|a_B\|_2 \le R^{-1/2} \quad \text{when} \quad B = \{x\,;\ |x - x_0| \le R\} \tag{8.20}$$

$$a_B(x) = 0 \quad \text{if} \quad x \notin B\,. \tag{8.21}$$

Then a function $f(x)$ belongs to F if and only if $f(x) = \sum_0^\infty \lambda_j a_{B_j}(x)$ where a_{B_j} satisfies (8.17) and (8.18) and

$$\sum_0^\infty |\lambda_j| < \infty\,. \tag{8.22}$$

The norm of f in F is the infimum of $\sum_0^\infty |\lambda_j|$ computed over all atomic decompositions of $f(x)$. We obviously have $\mathcal{S}(\mathbb{R}^3) \subset F$ and this continuous embedding is dense. moreover the dual space F^* coincides with the Morrey-Campanato space M^1.

In the opposite direction there exists a smallest Banach space which is adapted to the Navier-Stokes equations. This example will be revisited in section 24 and we now announce the following statement.

Lemma 8. *If E is adapted to the Navier-Stokes equations, then the homogeneous Besov space $\dot{B}_1^{2,1}(\mathbb{R}^3)$ is contained in E, this embedding is continuous and $\dot{B}_1^{2,1}(\mathbb{R}^3)$ is also adapted to the Navier-Stokes equations.*

The simplest description of $\dot{B}_1^{2,1}$ is the following one. If the orthonormal wavelet basis of section 6 is being used, then $f \in \dot{B}_1^{2,1}$ if and only if

$$f(x) = \sum_j \sum_k \alpha(j,k)\, 2^j \psi(2^j x - k)$$

with

$$\sum_j \sum_k |\alpha(j,k)| < \infty .$$

For the sake of simplicity, the sum over $\psi \in A$ (with $\sharp A = 7$) is ommited.

Before ending this section, let us state and prove an important lemma.

Lemma 9. *If E is adapted to the Navier-Stokes equations, for every φ in the Schwartz class $\mathcal{S}(\mathbb{R}^3)$, there exists a constant $C(\varphi)$ such that*

$$\|f * \varphi_t\|_\infty \le C(\varphi)\, t^{-1}\, \|f\|_E , \qquad t > 0 , \ f \in E , \tag{8.23}$$

with $\varphi_t(x) = t^{-3}\, \varphi(x/t)$.

Indeed $(f * \varphi_t)(x) = \int f(x + ty)\, \varphi(-y)\, dy$. We define a new function $f_{t,x}(y)$ by $f_{t,x}(y) = t\, f(ty + x)$ and observe that $\|f_{t,x}\|_E = \|f\|_E$. Moreover each function φ belonging to the Schwartz class yields a continuous linear form on E since E is continuously embedded in $\mathcal{S}'(\mathbb{R}^3)$. Therefore

$$| < f_{t,x}, \varphi > | \le C(\varphi)\, \|f_{t,x}\|_E = C(\varphi)\, \|f\|_E . \tag{8.24}$$

Now the scalar product $< f_{t,x}, \varphi >$ is precisely the integral we want to estimate, up to some trivial modifications.

The next section is a first step towards defining the concept of a mild solution of the Navier-Stokes equations. We begin with a simplified model which is the heat equation.

9 Mild solutions to the heat equation

We consider the heat equation with a forcing term $g = g(x,t)$. It reads

$$\begin{cases} \dfrac{\partial f}{\partial t} = \Delta f + g \\ f(x,0) = f_0(x) . \end{cases} \tag{9.1}$$

We are looking for a solution f of (9.1) as being continuous in the time variable with values in a functional Banach space E. For the time being, E is assumed to be separable and $f \in C([0,\infty); E)$ has the usual meaning.

Let us now consider a second functional Banach space E' with the following two properties

(9.2) $\Delta : E \to E'$ is continuous

(9.3) if φ belongs to the Schwartz class $\mathcal{S}(\mathbb{R}^3)$ then the convolution operator $f \to f * \varphi$ continuously maps E' into E

Definition 13. *With the preceding notations, a mild solution f to the heat equation* (9.1) *is a solution belonging to $C([0,\infty), E)$ when g belongs to $C([0,\infty), E')$.*

Let us denote by $S(t) = \exp(t\Delta)$, $t \geq 0$, the heat semi-group. We then have

Theorem 9.1. *If such a mild solution exists, it is uniquely given by*

$$f(x,t) = S(t)f_0 + \int_0^t S(t-s)\,g(s)\,ds\,. \tag{9.4}$$

This statement is trivial if $g = 0$ and $S(t)f_0$ obviously satisfies the required continuity with respect to the time variable if $f_0 \in E$. We can turn to the case where $f_0 = 0$. We then observe that the integral $\int_0^t S(t-s)\,g(s)\,ds$ belongs to $C([0,\infty), E')$. This integral does not belong to $C([0,\infty), E)$ in general and a counter-example is given at the end of this section. In other words, mild solutions do not exist in general when g belongs to $C([0,\infty), E')$.

It is interesting to have a sharper look at the integrand. A partial answer is given by the following lemma

Lemma 10. *If $\Delta : E \to E'$ is an isomorphism between the Banach spaces E and E', there exists a constant C such that*

$$\|S(t)f\|_E \leq Ct^{-1}\|f\|_{E'}\,. \tag{9.5}$$

For proving (9.5), one writes $f = \Delta u$, $u \in E$ with $\|u\|_E \leq C_0\|f\|_E$. Then $t\Delta S(t) = Q_t$ is a convolution operator with $\psi_t(x) = t^{-3}\psi(x/t)$ where ψ is the celebrated D. Marr's wavelet. Therefore $Q_t : E \to E$ is continuous and the operator norm does not depend on t.

Once (9.103) is obtained, we have $\|S(t-s)\,g(s)\|_E \le C'(t-s)^{-1}$ which diverges when s approaches t.

We now return to the proof of (9.102). We denote by φ a function in the Schwartz class whose Fourier transform $\hat{\varphi}(\xi)$ satisfies

$$\hat{\varphi}(\xi) = 1 \quad \text{on } |\xi| \le 1 \quad , \quad \hat{\varphi}(\xi) = 0 \quad \text{on } |\xi| \ge 2\,. \tag{9.6}$$

For $\varepsilon > 0$, we write $\varphi_\varepsilon(x) = \varepsilon^{-3}\,\varphi(x/\varepsilon)$ and denote by G_ε the convolution operator with φ_ε. We then obviously have

$$G_\varepsilon\, G_{2\varepsilon} = G_{2\varepsilon}\, G_\varepsilon = G_{2\varepsilon}\,. \tag{9.7}$$

Now $\frac{\partial f}{\partial t} = \Delta f + g$ implies

$$\frac{\partial f_\varepsilon}{\partial t} = A_\varepsilon f_\varepsilon + g_\varepsilon \tag{9.8}$$

where $f_\varepsilon = G_{2\varepsilon}(f)$, $g_\varepsilon = G_{2\varepsilon}(g)$ and $A_\varepsilon = \Delta G_\varepsilon$. This operator A_ε acts boundedly on E and (9.106) together with $f_\varepsilon(x,0) = 0$ obviously implies

$$f_\varepsilon(t) = \int_0^t \exp\big((t-s)A_\varepsilon\big)\, g_\varepsilon(s)\, ds\,. \tag{9.9}$$

The Fourier transform of the integrand reads

$$\exp\big(-(t-s)\,|\xi|^2\,\hat{\varphi}(\varepsilon\xi)\big)\,\hat{\varphi}(2\varepsilon\xi)\,\hat{g}(s,\xi)\,. \tag{9.10}$$

Since $\hat{\varphi}(\xi) = 1$ on the support of $\hat{\varphi}(2\xi)$, this can be simplified into

$$\exp\big(-(t-s)\,|\xi|^2\big)\,\hat{\varphi}(2\varepsilon\xi)\,\hat{g}(s,\xi) \tag{9.11}$$

and (9.107) can be rewritten as

$$f_\varepsilon(t) = \int_0^t s(t-s)\, g_\varepsilon(s)\, ds\,. \tag{9.12}$$

Finally $g_\varepsilon \to g$ in E' and $\|g_\varepsilon\|_{E'} \le C\|g\|_{E'}$. Lebesgue dominated convergence theorem implies the convergence in E' of f_ε to $\int_0^t S(t-s)\,g(s)\,ds$. But f_ε converges to f in E. It implies

$$f(t) = \int_0^t S(t-s)\, g(s)\, ds\,. \tag{9.13}$$

Indeed if a sequence u_j belongs to $E \cap E'$ and converges to u in E and to v in E', then for any function φ in the Schwartz class, $u_j * \varphi$ tends to $u * \varphi$ in E and also to $v * \varphi$ in E. It yields $u * \varphi = v * \varphi$ for every φ and $u = v$.

The unpleasant fact that mild solutions do not always exist in the framework of theorem is one of the main difficulties we will have to face for solving Navier-Stokes equations.

In the simplest case when $E' = L^2(\mathbb{R}^3)$ and E is the Sobolev space $H^2(\mathbb{R}^3)$, it is not true that $g \in C([0,\infty); L^2(\mathbb{R}^3))$ implies $f \in C([0,\infty); H^2(\mathbb{R}^3))$ as the most trivial counter-examples show.

There is however a simple and interesting situation where everything smoothly works.

Let us assume that $E = E_\alpha$, $\alpha > -3$, is defined by the following pointwise condition on the Fourier transform $\hat{f}(\xi)$ of f

$$|\hat{f}(\xi)| \leq C|\xi|^\alpha . \tag{9.14}$$

We then define $E' = E_{\alpha+2}$ and it is now a trivial check that $g \in C([0,\infty); E')$ implies $f \in C([0,\infty); E)$.

10 Mild solutions to Navier-Stokes equations : the L^3-case

Let us start with the simplest example which is provided by the adapted space $E = L^3(\mathbb{R}^3)$. We will then move to more involved cases which will culminate with the Morrey-Campanato space $M^2(\mathbb{R}^3)$.

Definition 14. *A mild solution of Navier-Stokes equations is a velocity field $u(x,t) \in C([0,\infty), L^3(\mathbb{R}^3))$ together with a pressure $p(x,t) \in C([0,\infty), L^{3/2}(\mathbb{R}^3))$ such that*

$$\begin{cases} \dfrac{\partial u}{\partial t} = \Delta u - \partial_j(u_j u) - \nabla p \\ \operatorname{div} u = 0 \\ u(x,0) = u_0(x) . \end{cases} \tag{10.1}$$

This equation needs to be given a meaning. First of all the Laplace operator Δ is an isomorphism from $E = L^3(\mathbb{R}^3)$ onto $E' = \{\Delta f\,;\, f \in L^3(\mathbb{R}^3)\}$. Such a statement is a tautology. But it is more interesting to observe that $\partial_j(u_j u) \in C([0,\infty); E')$ and so does ∇p.

Indeed $u \in C([0,\infty), L^3)$ implies $u_j u \in C([0,\infty); L^{3/2})$ and each function f in $L^{3/2}(\mathbb{R}^3)$ can uniquely be written as Λg where $\Lambda = (-\Delta)^{1/2}$ is the Calderón operator and $g \in L^3(\mathbb{R}^3)$. Applying this remark to $u_j u$ and p yields the required remark. Therefore $\frac{\partial u}{\partial t}$ belongs to $C([0,\infty); E')$.

Everything is now prepared to apply theorem 9.1. We write $w(x,t) = \partial_j(u_j u) + \nabla p$ and obtain

$$u(x,t) = S(t)u_0 + \int_0^t S(t-s)\, w(\cdot,s)\, ds\,. \tag{10.2}$$

This equation will be solved using Picard's fixed point theorem. However the main difficulty is the one we already mentioned in the linear case. Indeed the bilinear mapping which shows up in $\int_0^t S(t-s)\,w(\cdot,s)\,ds$ is not bounded for the natural norm we are using which is $\sup\{\|u(\cdot,t)\|_3\,;\, 0 \le t < \infty\}$.

This first example can be generalized. We begin with a definition

Definition 15. *Let E be a functional Banach space. We say that E is fully adapted to the Navier-Stokes equations if the following two properties are satisfied*

(10.3) *the Riesz transformations R_1, R_2 and R_3 act boundedly on E*

(10.4) *there exists a constant C such that for any two functions f and g belonging to E there exists a unique function h in E such that*

$$\Lambda h = fg \tag{10.4.a}$$

and

$$\|h\|_E \le C\|f\|_E\,\|g\|_E\,. \tag{10.4.b}$$

The bilinear operator defined by (10.4.a) is in some sense induced by the fundamental bilinear operator that governs the Navier-Stokes equations. This bilinear operator will be presented in Definition 18. A simpler scalar version is defined as

$$B(f,g) = \int_0^t \Lambda S(t-s)\, f(s)\, g(s)\, ds \tag{10.5}$$

where $f(s) = f(\cdot,s)$, $g(s) = g(\cdot,s)$ and $\Lambda = \sqrt{-\Delta}$.

We want to know whether B boundedly maps $C([0,\infty); E) \times C([0,\infty); E)$ into $C([0,\infty); E)$ when the norm in $C([0,\infty); E)$ is defined as

$$\sup_{t \ge 0} \|f(\cdot,t)\|_E\,. \tag{10.6}$$

We then have

Lemma 11. *If B boundedly maps $C([0,\infty);E) \times C([0,\infty);E)$ into $C([0,\infty);E)$, then* (10.4.a) *and* (10.4.b) *are satisfied.*

This is trivially checked. Indeed let $f(x,s) = f(x)$, $g(x,s) = g(x)$, $s \geq 0$, and let us compute $B(f,g)$. We obtain $B(f,g) = [I - S(t)]\,\Lambda^{-1}(fg)$ and $\|B(f,g)\|_E \leq C\|f\|_E\,\|g\|_E$ implies (10.4).

The converse statement is not true and the Lebesgue space $L^3(\mathbb{R}^3)$ is a counter-example. However the corresponding Lorentz space $L^{3,\infty}(\mathbb{R}^3)$ is an interesting example where B is bounded (theorem 18.2 of section 18).

The Besov spaces $\dot{B}_q^{-(1-3/q),\infty}(\mathbb{R}^3)$ are fully adapted to the Navier-Stokes equations as it will be proved in section 24. An other example is given by

$$E = \left\{ f \in \mathcal{S}'(\mathbb{R}^3)\,;\ |\hat{f}(\xi)| \leq C|\xi|^{-2} \right\}. \tag{10.7}$$

If f and g both belong to E, then $(f,g)\hat{}(\xi) = \hat{f} * \hat{g}$ satisfies $|(fg)\hat{}(\xi)| \leq C|\xi|^{-1}$ as a trivial computation shows. Then h is defined by $\hat{h}(\xi) = |\xi|^{-1}(fg)\hat{}(\xi)$ and we then have $|\hat{h}(\xi)| \leq C|\xi|^{-1}$.

An important observation is the following

Theorem 10.1. *The Morrey-Campanato space M^2 is not fully adapted to the Navier-Stokes equations.*

We begin with an obvious remark :

Lemma 12. *A function f of the two variables x_2 and x_3 belongs to $M^2(\mathbb{R}^3)$ if and only if $f(x_2,x_3) \in L^2(\mathbb{R}^2)$.*

Indeed one writes

$$\int_{|x|\leq R} |f(x_2,x_3)|^2\,dx \;\leq\; CR \tag{10.8}$$

and it suffices to let R tend to infinity to obtain the required equivalence.

If both f and g belong to M^2, then their pointwise product $f(x)\,g(x) = u(x)$ satisfies

$$\int_{B(x_0,R)} |u(x)|\,dx \;\leq\; C\,R \tag{10.9}$$

where C does not depend on x_0 or R.

If both f and g only depend on x_2 and x_3, so does u and $u(x_2, x_3)$ belongs to $L^1(\mathbb{R}^2)$.

If $u = \Lambda h$ in $\mathcal{S}'(\mathbb{R}^3)$, the same relation holds in $\mathcal{S}'(\mathbb{R}^2)$.

Finally u is an arbitrary function in $L^1(\mathbb{R}^2)$ and if $\iint u(x_2, x_3)\, dx_2\, dx_3 \neq 0$, then h cannot belong to $L^2(\mathbb{R}^2)$ since $\iint (\xi_2^2 + \xi_3^2)^{-1} |\hat{u}(\xi_2, \xi_3)|^2\, d\xi_2\, d\xi_3 = +\infty$.

We now return to the definition of mild solutions $u \in C([0,\infty); E)$ to the Navier-Stokes equations when E is a functional Banach space which is fully adapted to the Navier-Stokes equations.

This definition relies on the Leray projector we now want to define.

11 The Leray projection

The Leray projector is aimed to get rid of the pressure in the Navier-Stokes equations.

Keeping the same notations as above, let us denote by H the Hilbert space $(L^2(\mathbb{R}^3))^3$ and by $H_0 \subset H$ the closed linear subspace defined by $\partial_1 u_1 + \partial_2 u_2 + \partial_3 u_3 = 0$. This condition will be rewritten as

$$(11.1) \qquad R_1 u_1 + R_2 u_2 + R_3 u_3 = 0$$

where R_1, R_2 and R_3 are the Riesz transformations.

The Leray projector $\mathbb{P}$ is defined as the orthogonal projection from H onto H_0.

The computation of $\mathbb{P}$ is straightforward and one obtains

$$(11.2) \qquad \mathbb{P}\begin{pmatrix} u_1 \\ u_2 \\ u_3 \end{pmatrix} = \begin{pmatrix} u_1 - R_1(\sigma) \\ u_2 - R_2(\sigma) \\ u_3 - R_3(\sigma) \end{pmatrix}$$

where

$$(11.3) \qquad \sigma = R_1 u_1 + R_2 u_2 + R_3 u_3 .$$

The Leray projector $\mathbb{P}$ is acting on $(L^2(\mathbb{R}^3))^3$ but this action can be extended to every functional Banach space on which the Riesz transformations act boundedly.

12 Mild solutions of Navier-Stokes equations : the general case

We now consider a Banach space E which is fully adapted to Navier-Stokes equations and define the two functional spaces E' and Π by the following conditions :

(12.1) $\Delta : E \to E'$ is an isometrical isomorphism

(12.2) $\Lambda : E \to \Pi$ is an isometrical isomorphism.

It will be proved (lemma 13) that the kernel of Δ is reduced to $\{0\}$ and that the same property holds for Λ.

Definition 16. *Let the functional Banach space E be fully adapted to the Navier-Stokes equations. Let $E' = \Delta E$ and $\Pi = \Lambda E$ as above.*

A mild solution solution $\big(u(x,t), p(x,t)\big)$ to the Navier-Stokes equations is defined by the following conditions

(12.3) $u(x,t) \in C([0,\infty); E)$

(12.4) $\dfrac{\partial u}{\partial t}(x,t) \in C([0,\infty); E')$

(12.5) $p(x,t) \in C([0,\infty); \Pi)$

(12.6) $\dfrac{\partial u}{\partial t} = \Delta u - \partial_j(u_j u) - \nabla p$

(12.7) $\operatorname{div} u = 0$

(12.8) $u(x,0) = u_0(x)\,.$

Let us make a few remarks about this definition.

If E is a separable functional Banach space, then $C([0,\infty); E)$ has the usual meaning. Otherwise E is assumed to be the dual space F^* of a separable Banach space F and $u \in C([0,\infty); E)$ always means

(12.9) $u \in C((0,\infty); E)$ when E is given its norm topology

(12.10) $u \in C([0,\infty); E)$ when E is given its weak-star topology.

Concerning the non-linearity in the Navier-Stokes equations, we have $u_j u_k = \sum_1^3 \partial_\ell(v_{j,k,\ell})$ by (10.4). Then $v_{j,k,\ell}$ belong to E and $\sum_1^3 \partial_j(u_j u_k) = \sum_1^3 \sum_1^3 \partial_j \partial_\ell(v_{j,k,\ell})$ belongs to E'. Similarly Δu and ∇p belong to E'.

These remarks lead to the following

Lemma 13. *If the functional Banach space E is fully adapted to the Navier-Stokes equations, then each of the terms which appear in the left-hand side or in the right-hand side of* (12.130) *belongs to* $C([0,\infty);E')$.

We now reach our main goal

Theorem 12.1. *If $\big(u(x,t),p(x,t)\big)$ is a mild solution of the Navier-Stokes equations, we then have*

$$\frac{\partial u}{\partial t} = \Delta u - \mathbb{P}\big(\partial_j(u_j u)\big) \tag{12.11}$$

while the pressure $p(x,t)$ is given by

$$p = R_j\, R_k(u_j u_k)\,. \tag{12.12}$$

If conversely $u \in C([0,\infty);E)$, $\frac{\partial u}{\partial t} \in C([0,\infty);E')$ and $\operatorname{div} u_0(x) = 0$, then (12.135) *and* (12.136) *imply that $\big(u(x,t),p(x,t)\big)$ is a mild solution to the Navier-Stokes equations.*

This fundamental theorem is an easy consequence of the following remark.

Lemma 14. *If v belongs to F and satisfies $\mathbb{P}(v) = 0$, then $v = \nabla p$ where p is a scalar function and p belongs to Π. Moreover p is uniquely defined by this equation.*

For proving this remark, we return to the definition of $\mathbb{P}$. Then $\mathbb{P}(v) = 0$ reads $v_1 = R_1(\sigma)$, $v_2 = R_2(\sigma)$, $v_3 = R_3(\sigma)$ where $\sigma = R_1v_1 + R_2v_2 + R_3v_3$. We have $\sigma \in E'$ since the Riesz transformations act boundedly on E'. Since $\Lambda : \Pi \to E'$ is onto, σ can be written $\sigma = \Lambda p$ and this is the required conclusion.

For proving that $\Lambda : \Pi \to E'$ is $1-1$, we assume $\Lambda p = 0$. Since p belongs to the range of Λ, we write $p = \Lambda f$ and obtain $\Delta f = 0$. Therefore f is a harmonic function. Since E is contained in $\mathcal{S}'(\mathbb{R}^3)$, f is a harmonic polynomial. Let φ be a function in the Schwartz class and $\varphi_t(x) = t^{-3}\,\varphi(x/t)$.

Then $\|f * \varphi_t\|_\infty \le Ct^{-1}$ (lemma 10). But $(f*\varphi_t)(x) = \int f(x-yt)\,\varphi(y)\,dy$ is a polynomial in t. Therefore this polynomial identically vanishes and $f = 0$.

We now return to theorem 12.1. If (u,p) is a mild solution, we have $\mathbb{P}(u) = u$, $\mathbb{P}(\Delta u) = \Delta u$, $\mathbb{P}(\frac{\partial u}{\partial t}) = \frac{\partial u}{\partial t}$ and finally $\mathbb{P}(\nabla p) = 0$. These identities yield (12.135). Moreover applying the divergence operator to (12.130)

identities yield (12.11). Moreover applying the divergence operator to (12.6) we obtain $\Delta p + \partial_j \partial_k (u_j u_k) = 0$. But the functional space Π cannot contain polynomials.

If p is such a polynomial, we have $p = \Lambda f$ where $f \in E$. Returning to the Fourier transforms, we obtain $|\xi|\, \hat{f}(\xi) = 0$ on $\mathbb{R}^3 \setminus \{0\}$. Therefore f is a polynomial. Since $f \in E$, we have $f = 0$. We then obtain $p = R_j R_k (u_j u_k)$ as announced.

Let us once more stress that $u_j u_k = \sum_1^3 \partial_\ell (v_{j,k,\ell})$ where $v_{j,k,\ell} \in E$ which yields $p \in \Pi$.

We now prove the converse statement in theorem 12.1. We begin with a simple observation.

Lemma 15. *If* $u(x,t) \in C([0,\infty); E)$, $\operatorname{div} u_0 = 0$ *and* (12.11) *is satisfied, then* $\operatorname{div} u(\cdot, t) = 0$ *for* $t \geq 0$.

For proving lemma 15, we apply the divergence operator to (12.11) and obtain $\frac{\partial w}{\partial t} = \Delta w$ where $w = \operatorname{div} u$. We have $w \in C([0,\infty), \Pi)$ and $w(\cdot, 0) = \operatorname{div} u_0 = 0$.

Theorem 9.1 applies and yields $w = 0$. Next we consider the difference

$$\varepsilon(x,t) = \frac{\partial u}{\partial t} - \Delta u + \partial_j (u_j u)\,. \tag{12.13}$$

Then (12.11) and $\operatorname{div} u = 0$ imply $\mathbb{P}[\varepsilon(\cdot, t)] = 0$ in $C([0,\infty); E')$. This yields $\varepsilon(\cdot, t) = -\nabla p$ and $p \in C([0,\infty), \Pi)$. Our goal is now achieved and $\big(u(x,t), p(x,t)\big)$ is the mild solution we looking for.

This proof paves the way to our second definition of a mild solution. In this definition, the pressure has disappeared.

Definition 17. *Let* E *be a Banach space which is fully adapted to the Navier-Stokes equations. Let us assume as we did above that* $\Delta : E \to E'$ *is an isometrical isomorphism. Then a mild solution to the Navier-Stokes equations is defined by the following three conditions*

$$u \in C([0,\infty); E) \tag{12.14}$$

$$\frac{\partial u}{\partial t} \in C([0,\infty); E') \tag{12.15}$$

$$\frac{\partial u}{\partial t} = \Delta u - \mathbb{P}\big(\partial_j (u_j u)\big)\,. \tag{12.16}$$

13 More about mild solutions

A fundamental observation concerning definition 17 is the following : if $u(x,t)$ is a mild solution to the Navier-Stokes equations, the same is true for $u(x,t+\tau)$, $\tau \geq 0$.

This remark will be used in section 20.

We now want to drop the condition (12.139) in the definition of a mild solution to the Navier-Stokes equations.

Following T. Kato, we apply theorem 9.1 to (12.140) and reach the third definition of a mild solution

Definition 18. *With the same notations as above, a mild solution of the Navier-Stokes equations is defined by the following two conditions*

$$(13.1) \qquad u(\cdot,t) \in C([0,\infty);E)$$

$$(13.2) \qquad u(\cdot,t) = S(t)u_0 - \mathbb{P}\int_0^t S(t-s)\,[\partial_j(u_j u)(s)]\,ds\,.$$

As was already mentioned when theorem 9.1 was proved, the status of this integral is far from being obvious. Indeed $\partial_j(u_j u)$ belongs to $C([0,\infty);E')$ and our integral is an E' valued Bochner integral. However it is not true that this integral should belong to E. If $u(\cdot,t)$ is a mild solution to the Navier-Stokes equations, it is the case and our integral belongs to E. What we want to stress is the following fact. There is no hope to solve (13.142) by a genuine application of Picard's fixed point theorem.

This issue will be fixed by a remarkable approach dued to T. Kato and his collaborators. This approach will be presented in section 19. For the time being, we will impose a digression to the reader. This digression is aimed to study the paraproduct algorithm. This algorithm is needed to proving that $\dot{B}_q^{-(1-3/q),\infty}$ is fully adapted to the Navier-Stokes equations.

As was already stated, the Morrey-Campanato space M^2 is not fully adapted to the Navier-Stokes equations. Indeed (10.116) is not satisfied while the Riesz transformations are bounded on M^2.

For defining the concept of a mild solution to the Navier-Stokes equations when $E = M^2$, we return to theorem 9.1 and define a suitable functional Banach space E'. This Banach space should contain all functions or distributions of the form Δf where $f \in M^2$. Then the mapping $\Delta : E \to E'$ will be continuous.

This Banach space should also contains all functions of the form $\partial_1 u_1 + \partial_2 u_2 + \partial_3 u_3$ when u_1, u_2 and u_3 satisfy the estimate $\sup_{x_0} \sup_{0<R<\infty} \frac{1}{R} \cdot \int_{|x-x_0|\leq R} |f(x)|\, dx < \infty$. Finally E' should be invariant under the Riesz transformation.

The explicit construction of E' is not trivial but will not be needed in what follows.

Right now a digression is needed. Apart from two trivial examples of fully adapted Banach spaces, all the interesting examples are based upon the paraproduct algorithm. This algorithm is also playing a crucial role in the proof of uniqueness when $E = L^3(\mathbb{R}^3)$. The next section contains a review on the paraproduct algorithms.

14 The paraproduct algorithm

There are two versions of the paraproduct algorithm. The first one is implicit in [20] and was given its full strength and flexibility by J.M. Bony. The second one is a wavelet based algorithm and is related to P. Federbush's program.

The Bony's paraproduct is using the Littlewood-Paley analysis.

For defining a Littlewood-Paley analysis we fix a function $\varphi(x)$ in the Schwartz class $\mathcal{S}(\mathbb{R}^3)$ such that $\hat{\varphi}(\xi) = 1$ on the ball $|\xi| \leq 1/2$ while $\hat{\varphi}(\xi) = 0$ outside the ball $|\xi| < 1$.

We also may assume $0 \leq \hat{\varphi}(\xi) \leq 1$ everywhere and that φ (or $\hat{\varphi}$) is a radial function. But these two properties are not needed in what follows.

For $j \in \mathbf{Z}$, we write $\varphi_j(x) = 2^{3j}\, \varphi(2^j x)$ and we denote by S_j the convolution operator defined by $S_j(f) = f * \varphi_j$.

If E is a functional Banach space, we have $\lim_{j\to+\infty} \|S_j(f) - f\|_E = 0$ when $\mathcal{S}(\mathbb{R}^3)$ is dense in E. If $E = F^*$ where F is such a functional Banach space, we have $S_j(f) \rightharpoonup f$ $(j \to +\infty)$ where here the limit exists for the $\sigma(E, F)$ topology.

A functional Banach space E is adapted to the Littlewood-Paley decomposition if

$$\lim_{j\to-\infty} \|S_j(f)\|_E = 0 \tag{14.1}$$

for every $f \in E$. This condition concerns the case where $\mathcal{S}(\mathbb{R}^3)$ is dense in E. If $E = F^*$, we instead impose $S_j(f) \rightharpoonup 0$ in $\sigma(E, F)$ as $j \to -\infty$.

We next write $\Delta_j = S_{j+1} - S_j$. If E is adapted to the Littlewood-Paley decomposition, we have

(14.2) $$f = \sum_{-\infty}^{\infty} \Delta_j(f)$$

where this series converges to f in E when $\mathcal{S}(\mathbb{R}^3)$ is dense in E and converges to f in $\sigma(E, F)$ when $E = F^*$.

Among adapted Banach spaces, one lists the Lebesgue spaces $L^p(\mathbb{R}^3)$ when $1 < p < \infty$. However L^1 is not adapted. Indeed if $f \in L^1$ and $\int f\,dx = 1$, then the right-hand side of (14.144) cannot converge to f for this L^1-norm since $\int \Delta_j(f)\,dx = 0$.

Similarly L^∞ is not adapted to the Littlewood-Paley decomposition. An obvious counter-example is given by $f = 1$ where $\Delta_j(f) = 0$ for all j's.

We have $\Delta_j(f) = f * \psi_j$ where $\psi_j(x) = 2^{3j}\,\psi(2^j x)$ and $\psi(x) = 8\varphi(2x) - \varphi(x)$. The Fourier transform $\hat{\psi}(\xi)$ of ψ is compactly supported and vanishes on $|\xi| \leq 1/2$ and outside the ball $|\xi| \leq 2$. Therefore ψ is similar to a mother wavelet and a Littlewood-Paley analysis is similar to a wavelet analysis [55].

The paraproduct algorithm was first designed for studying some involved bi-linear operators $B(f,g)$ which appear in Calderón's program [56]. Later on J.M. Bony observed that this algorithm was also relevant for studying the ordinary product $f(x)\,g(x)$ whenever this product cannot be given an ordinary meaning. This situation occurs when f and g are both badly behaved (e.g. $f = g = \delta_0$ where δ_0 is the Dirac mass at 0). In this situation the paraproduct algorithm yields an "additive renormalization" or "regularization" of a divergent expansion.

Nowadays para-product algorithms are expected to play a role in numerical analysis [54].

It is now time to unveil the paraproduct algorithm which reads

(14.3) $$f(x)\,g(x) = \sum_{-\infty}^{\infty} A_j(x) + \sum_{-\infty}^{\infty} B_j(x) + \sum_{-\infty}^{\infty} C_j(x)$$

where

(14.4) $$A_j(x) = S_{j-2}(f)\,\Delta_j(g) \quad , \quad B_j(x) = S_{j-2}(g)\,\Delta_j(f)$$

while

$$C_j(x) = \big(\Delta_{j-2}(f) + \cdots + \Delta_{j+2}(f)\big)\,\Delta_j(g)\,. \tag{14.5}$$

The relevance of this splitting of the product $f(x)\,g(x)$ will be clarified by the following remark.

Lemma 16. *Let A and B be two compact sets in $\mathbb{R}^3$ and $A+B$ denote the compact set of all sums $\xi+\eta$, $\xi \in A$, $\eta \in B$.*

If $f(x)$ and $g(x)$ are two functions such that the Fourier transform $\hat{f}$ of f is supported by A and the Fourier transform $\hat{g}$ of g is supported by B, then the Fourier transform of the pointwise product $f(x)\,g(x)$ is supported by $A+B$.

If this obvious remark is applied to $S_{j-2}(f)\,\Delta_j(g) = A_j$, we find that $\hat{A}_j(\xi)$ is supported by the dyadic annulus $\frac{1}{4}\,2^j \le |\xi| \le \frac{9}{4}\,2^j$. Therefore the series $\sum_{-\infty}^{\infty} A_j(x)$ is behaving like a Littlewood-Paley series. The same applies to $\sum_{-\infty}^{\infty} B_j(x)$. If, for instance, both f and g belong to $L^1(\mathbb{R}^3)$, then $\sum_{-\infty}^{\infty} A_j(x)$ and $\sum_{-\infty}^{\infty} B_j(x)$ are convergent in the Besov space $\dot{B}_1^{-3,\infty}$. We will return to this point.

The devil is hidden in the third series $\sum_{-\infty}^{\infty} C_j(x)$. Indeed $C_j(x)$ contains the square terms $\Delta_j(f)\,\Delta_j(g)$ and a few other terms. If, for example, $g = \overline{f}$ and if φ is real valued, then $\Delta_j(f)\,\Delta_j(g) = |\Delta_j(f)|^2$ and there are no cancellations. If both f and g belong to $L^1(\mathbb{R}^3)$ the series $\sum_{-\infty}^{\infty} C_j(x)$ diverges. Indeed one cannot multiply two L^1 functions. Using the paraproduct algorithm, we isolate the few terms which are responsible for this issue.

Using a metaphore, the paraproduct algorithm can be compared to the celebrated Richardson's cascade. Indeed high frequencies cascade to low frequencies in each product $\Delta_j(f)\,\Delta_j(g)$.

There is a second version of the paraproduct algorithm which is based on a wavelet series expansion.

Let ψ_λ, $\lambda \in \Lambda$, be the orthonormal wavelet basis which is constructed in section 6. We have $\Lambda = \mathbf{Z} \times \mathbf{Z}^3 \times E$ where E is the finite set of $\varepsilon = (\varepsilon_1, \varepsilon_2, \varepsilon_3) \in \{0,1\}^3$ with $\varepsilon \ne (0,0,0)$. if $\lambda(j,k,\varepsilon)$, $j \in \mathbf{Z}$, $k \in \mathbf{Z}^3$, $\varepsilon \in E$, we then have $\psi_\lambda(x) = 2^{3j/2}\,\psi_\varepsilon(2^j x - k)$. Finally $\psi_\varepsilon(x)$ belongs to the Schwartz class $\mathcal{S}(\mathbb{R}^3)$ and $\hat{\psi}_\varepsilon(\xi) = 0$ if either $|\xi| \le 2\pi/3$ or $|\xi| \ge 8\pi/3$. Here $|\xi| = \sup\big(|\xi_1|, |\xi_2|, |\xi_3|\big)$.

We consider two functions f and g in $L^2(\mathbb{R}^3)$. Then $f(x) = \sum \alpha_\lambda \psi_\lambda(x)$ and $g(x) = \sum \beta_\lambda \psi_\lambda(x)$. This leads to

$$
\begin{aligned}
(14.6) \qquad f(x)\,g(x) &= \sum_{\lambda\in\Lambda} \alpha_\lambda \beta_\lambda \psi_\lambda^2(x) + \sum\sum_{\lambda\neq\lambda'} \alpha_\lambda \beta_{\lambda'} \psi_\lambda(x) \psi_{\lambda'}(x) \\
&= U(x) + V(x)\,.
\end{aligned}
$$

If for simplifying the discussion, ψ_E, $\varepsilon \in E$, are real valued, the square terms $\psi_\lambda^2(x)$ satisfy $\int \psi_\lambda^2(x)\,dx = 1$. In contrast the rectangle terms satisfy $\int \psi_\lambda(x)\,\psi_{\lambda'}(x)\,dx = 0$. Therefore the series $U(x)$ is likely to diverge while the series $V(x)$ is like to converge. This convergence is provided by the cancellations which are given by the orthogonality.

In order to extend the definition of the product $f(x)\,g(x)$ to some functional settings where this product has no meaning, it suffices to eliminate the divergent series $\sum \alpha_\lambda\, \beta_\lambda\, \psi_\lambda^2(x)$.

However there are many examples where the cancellation provided by $\int \psi_\lambda(x)\,\psi_{\lambda'}(x)\,dx = 0$ is not sufficient. In order to prove the convergence of $V(x)$ we often need a stronger cancellation such as $\int x^\alpha\,\psi_\lambda(x)\,\psi_{\lambda'}(x)\,dx = 0$ for $|\alpha| \leq N$ where the integer N is related to the functional setting. But orthonormal wavelet bases do not have this strong orthogonality. That is why the wavelet based paraproduct was given up [54].

15 Examples of Banach spaces which are fully adapted to the Navier-Stokes equations

We first list two trivial examples for which the paraproduct algorithm is not needed.

The first one is the Lebesgue space $L^3(\mathbb{R}^3)$. Its role was discovered by T. Kato [43] and we will return to the uniqueness issue in Kato's theorem in section 20.

For proving that $L^3(\mathbb{R}^3)$ is fully adapted we concentrate on (10.116). Indeed singular integral operators are bounded on $L^3(\mathbb{R}^3)$.

If both $f(x)$ and $g(x)$ belong to $L^3(\mathbb{R}^3)$, then $f(x)\,g(x)$ belongs to $L^{3/2}(\mathbb{R}^3)$ and $h = \Lambda^{-1}(fg)$ is back in $L^3(\mathbb{R}^3)$ as Sobolev embedding theorem tells us.

The same approach is valid for the Lorentz space $L^{3,\infty}(\mathbb{R}^3)$. Instead of writing the proof, let us concentrate on an interesting example. The function $f(x) = |x|^{-1}$ obviously belongs to $L^{3,\infty}(\mathbb{R}^3)$ and the special case $f(x) = g(x) = |x|^{-1}$ yields $h(x) = c|x|^{-1}$ where c is a constant.

This example will introduce the third example where E is defined by a pointwise estimate on the Fourier transform $\hat{f}(\xi)$ of f. This estimate reads

$$|\hat{f}(\xi)| \leq C|\xi|^{-2}. \tag{15.1}$$

If both f and g belong to E, then

$$|(fg)\hat{}(\xi)| \leq |\hat{f}| * |\hat{g}| \leq C|\xi|^{-1}$$

which obviously implies $fg = \Lambda(h)$ with $h \in E$.

The next examples we want to treat are the Besov spaces $B_q = \dot{B}_q^{\alpha,\infty}$, $\alpha = 3/q - 1$. The paraproduct algorithm will be used to show that B_q is fully adapted to the Navier-Stokes equations when $1 \leq q < 3$.

The discussion will be clarified if the general definition of homogeneous Besov spaces is presented.

Definition 19. *If $1 \leq p \leq \infty$, $1 \leq q \leq \infty$ and if s is a real number, the homogeneous Besov space $\dot{B}_p^{s,q}(\mathbb{R}^n)$ is defined by*

$$\|\Delta_j(f)\|_p \leq \varepsilon_j 2^{-js} \qquad j \in \mathbf{Z} \tag{15.2}$$

where

$$\left(\sum_{-\infty}^{\infty} \varepsilon_j^q\right)^{1/q} < \infty. \tag{15.3}$$

This definition should be completed with

$$\|S_j(f)\|_\infty \leq \bar{\varepsilon}_j 2^{-j(s-n/p)} \tag{15.4}$$

if $s < n/p$. We also impose

$$\left(\sum_{-\infty}^{\infty} \bar{\varepsilon}_j^q\right)^{1/q} < \infty. \tag{15.5}$$

Let us explain why (15.4) is a natural *a priori* assumption on a function f in $\dot{B}_p^{s,q}$.

For this purpose, the celebrated Bernstein's inequalities will be used. They are given in a slightly more general context.

Lemma 17. *There exists a constant C such that, for $1 \le p \le q \le \infty$, one has*

$$\|f\|_q \le C\, R^{n(1/p-1/q)}\, \|f\|_p \tag{15.6}$$

whenever f belongs to $L^p(\mathbb{R})$ and its Fourier transform $\hat{f}(\xi)$ is supported by $|\xi| \le R$.

Returning to (15.152) we have

$$\|\Delta_j(f)\|_\infty \le C\, \varepsilon_j\, 2^{-j(s-n/p)}\,. \tag{15.7}$$

If the Littlewood-Paley expansion of f converges to f in the distributional sense, we can write

$$S_j(f) = \sum_{j'<j} \Delta_{j'}(f) \tag{15.8}$$

and (15.155) clearly implies (15.152) and $\overline{\varepsilon}_j \in \ell^q(\mathbf{Z})$.

If conversely (15.152) holds, then $\|S_j(f)\|_\infty \to 0$ $(j \to -\infty)$ and the Littlewood-Paley expansion of f converges to f. This means that (15.152) can be replaced by the seemingly weaker condition

$$S_j(f) \rightharpoonup 0 \qquad \text{in} \quad \sigma(\mathcal{S}', \mathcal{S})\,. \tag{15.9}$$

We now concentrate on the specific homogeneous Besov space $\dot{B}_q^{\alpha,\infty}$ where $\alpha = 3/q - 1$. It is defined by

$$\|\Delta_j(f)\|_q \le C\, 2^{-j\alpha}\,, \quad j \in \mathbf{Z}\,,\ \alpha = 3/q - 1 \tag{15.10}$$

together with

$$\|S_j(f)\|_\infty \le C\, 2^j\,. \tag{15.11}$$

We then have

Theorem 15.1. *The Banach space B_q is fully adapted to the Navier-Stokes equations if $1 \le q < 3$.*

Let us begin with the dilation invariance of the norm. The norm of f in B_q is defined as

$$\|f\| = \sup_{j \in \mathbf{Z}} 2^{j(3/q-1)}\, \|\Delta_j(f)\|_q \tag{15.12}$$

which obviously implies

$$\|2^m f(2^m x)\| = \|f\| \quad \text{if} \quad m \in \mathbf{Z}. \tag{15.13}$$

On the other hand there exist two constants $C_2 \geq C_1 > 0$ such that

$$C_1\|f\| \leq \|f(\lambda x)\| \leq C_2\|f\| \tag{15.14}$$

if $1 \leq \lambda \leq 2$.

We then introduce a new norm which is defined by

$$\sup\{\lambda\|f(\lambda x)\|\,;\, 0 < \lambda < \infty\} = |||f|||. \tag{15.15}$$

This new norm satisfies

$$C_1\|f\| \leq |||f||| \leq C_2\|f\|\,, \qquad f \in B_q \tag{15.16}$$

and we have

$$|||f(x)||| = |||\lambda f(\lambda x)|||\,, \qquad 0 < \lambda < \infty. \tag{15.17}$$

This being checked, we turn to the difficult statement in theorem 15.1. It is based on the following lemma

Lemma 18. *Let f_j, $j \in \mathbf{Z}$, be a sequence of functions in $L^p(\mathbb{R}^n)$. Let us assume the following two conditions*

(15.18) *the Fourier transform $\hat{f}_j(\xi)$ of f_j is supported by the ball $|\xi| \leq 2^j$*

$$\|f_j\|_p \leq \varepsilon_j 2^{-js} \quad \textit{where} \quad s > 0 \quad \textit{and} \quad \sum_{-\infty}^{\infty} \varepsilon_j^q < \infty. \tag{15.19}$$

Then $\sum_{-\infty}^{\infty} f_j$ belongs to $\dot{B}_p^{s,q}$.

Similarly if the Fourier transform $\hat{f}_j(\xi)$ of f_j vanishes on the ball $|\xi| \leq 2^j$ and if (15.19) holds with $s < 0$ and $\sum_{-\infty}^{\infty} \varepsilon_j^q < \infty$, then $\sum_{-\infty}^{\infty} f_j$ belongs to $\dot{B}_p^{s,q}$.

Let us prove this statement when $s > 0$. We have $\Delta_m(\sum_{-\infty}^{\infty} f_j) = \Delta_m(\sum_{m-2}^{\infty} f_j) = w_m$ by (15.18). Then

$$\|w_m\|_p \leq C\sum_{m-2}^{\infty}\|f_j\|_p \leq C\sum_{m-2}^{\infty}\varepsilon_j 2^{-js} = \eta_m 2^{-ms} \quad \text{where} \quad \eta_m \in \ell^q(\mathbf{Z}).$$

It should be observed that (15.18) can be replaced by a much weaker statement. If indeed $0 < s < m$, $m \in \mathbb{N}$, it suffices to assume the following

$$\|\partial^\alpha f_j\|_p \le \eta_j\, 2^{j(m-s)}\,, \qquad |\alpha| = m \tag{15.20}$$

with

$$\sum_{-\infty}^{\infty} \eta_j^q < \infty\,. \tag{15.21}$$

Then $\sum_{-\infty}^{\infty} f_j$ belongs to $\dot{B}_p^{s,q}$.

We now return to $B^q = \dot{B}_q^{-(1-3/q),\infty}$ when $1 \le q < 3$. Writing $f = S_0(f) + \Delta_0(f) + \cdots + \Delta_j(f) + \cdots$ we have $S_0(f) \in L^\infty(\mathbb{R}^3)$ while $\|\Delta_j(f)\|_q \le \varepsilon_j\, 2^{j(1-3/q)}$. If $1 \le q \le 2$, this implies $\|\Delta_j(f)\|_2 \le \varepsilon_j\, 2^{-j/2}$ by Bernstein's inequalities and $\sum_0^\infty \Delta_j(f) \in L^2$. If $2 < q < 3$, we obviously have $\sum_0^\infty \Delta_j(f) \in L^q$. Finally f always belongs to L^2_{loc}.

For proving (10.4) the paraproduct algorithm will be used. Then $f(x)\,g(x) = A(x) + B(x) + C(x)$ where

$$A(x) = \sum_{-\infty}^{\infty} S_{j-2}(f)\,\Delta_j(g) \tag{15.22}$$

$$B(x) = \sum_{-\infty}^{\infty} S_{j-2}(g)\,\Delta_j(f) \tag{15.23}$$

and

$$C(x) = \sum\sum_{|j'-j|\le 2} \Delta_j(g)\,\Delta_{j'}(f)\,. \tag{15.24}$$

Both A and B are Littlewood-Paley expansions. We observe that $\|S_{j-2}(f)\cdot \Delta_j(g)\|_q \le C\,2^{j(2-3/q)}$ and therefore obtain $A \in \dot{B}_q^{\beta,\infty}$ where $\beta = \alpha - 1$. Therefore $A = \Lambda a$ with $a \in \dot{B}_q^{\alpha,\infty} = B_q$. The same remark applies to B.

As always in the paraproduct algorithm, the devil is hidden inside the third series $C(x)$.

Indeed we have $\|\Delta_j(g)\,\Delta_{j'}(f)\|_q \le C\,2^{j(2-3/q)}$ which only permits to apply lemma 18 if $1 \le q < 3/2$. If $3/2 \le q < 3$, the idea consists in fixing r such that $\sup(2,q) \le r < 3$ and in applying Bernstein's inequalities. We obtain

$$\|\Delta_j(f)\|_r \le C\,2^{-3\varepsilon j} \qquad \text{where } \frac{1}{r} = \frac{1}{3} + \varepsilon\,, \quad \varepsilon > 0\,,$$

which yields $\|\Delta_j(f)\,\Delta_{j'}(g)\|_{r/2} \le C\,2^{-6\varepsilon j}$.

Now lemma 18 can be applied and we obtain $C \in \dot{B}_{r/2}^{6\varepsilon,\infty}$. Therefore $\|\Delta_j(C)\|_{r/2} \le C\,2^{-6\varepsilon j}$ which implies by Bernstein's inequalities $\|\Delta_j(C)\|_q \le C\,2^{j(2-3/q)}$ and the proof ends as before.

When $q = 1$, $B_q = \dot{B}_1^{2,\infty}$. The proof we just gave works as well for the space $\dot{B}_1^{2,1}$ we want to study now.

This Banach space has a fascinating property. Among all the functional Banach spaces E which are adapted to the Navier-Stokes equations, $\dot{B}_1^{2,1}$ is the smallest one. This minimality property is easily explained. Let us consider the orthonormal wavelet basis $2^{3j/2}\,\psi(2^j x - k)$, $j \in \mathbf{Z}$, $k \in \mathbf{Z}^3$, $\psi \in A$ (with $\sharp\, A = 7$, $A \subset \mathcal{S}(\mathbb{R}^3)$).

We then have $\|2^j\psi(2^j x - k)\|_E = \|\psi\|_E$. Since E is a Banach space, this implies $\sum\sum \alpha(j,k)\,2^j\psi(2^j x - k) \in E$ whenever $\sum\sum|\alpha(j,k)| \le 1$. But any $f \in \dot{B}_1^{2,1}$ is given by a wavelet expansion $\sum\sum \alpha(j,k)\,2^j\psi(2^j x - k)$ where $\sum\sum|\alpha(j,k)| < \infty$. This implies the minimality of $\dot{B}_1^{2,1}$.

There is a second approach to $\dot{B}_1^{2,1}$ which relates this Banach space to the bump algebra.

The bump algebra is defined by the following properties : if $g_{(\alpha,a)}(x) = \exp(-\alpha|x-a|^2)$, $\alpha > 0$, $a \in \mathbb{R}^3$, then a function in the bump algebra can be written as a series

$$f(x) = \sum_0^\infty \lambda_j\, g_{(\alpha_j,a_j)}(x) \tag{15.25}$$

where $\alpha_j > 0$, $a_j \in \mathbb{R}^3$ and $\sum_0^\infty |\lambda_j| < \infty$.

Such an expansion is not unique and the norm of $f(x)$ in the bump algebra is defined as the infimum of $\sum_0^\infty |\lambda_j|$ over all expansions of $f(x)$. Finally the bump algebra is closed under pointwise multiplication. This is obvious since $g_{(\alpha,a)}\,g_{(\beta,b)} = \lambda g_{(\gamma,c)}$ where $0 < \lambda \le 1$, $\gamma = \alpha + \beta$ and $c = (\alpha a + \beta b)/\alpha + \beta$.

The bump algebra is identical to the Besov space $\dot{B}_1^{3,1}$ and each $f \in \dot{B}_1^{2,1}$ can be written as $f = \Lambda f$ where $g \in \dot{B}_1^{3,1}$.

Before ending this section, a variant on the preceding proof will be given. The following result will be needed in the proof of theorem 26.1.

Theorem 15.2. *If* $3/2 < q < 6$, $f \in L^q(\mathbb{R}^3)$ *and* $g \in \dot{B}_2^{1/2,\infty}$, *then* *the pointwise product* $f(x)\,g(x)$ *belongs to the homogeneous Besov space* $\dot{B}_2^{1/2-3/q,\infty}$.

For proving this fact, the paraproduct algorithm is again used.

The first series is $A(x) = \sum_{-\infty}^{\infty} A_j(x)$ where $A_j = S_{j-2}(f)\,\Delta_j(g)$. Since f belongs to $L^q(\mathbb{R}^3)$, we have

$$\|S_{j-2}(f)\|_\infty \leq C\,2^{3j/q}\,\|f\|_q \tag{15.26}$$

by Bernstein's inequalities. This yields

$$\|A_j\|_2 \leq C\,2^{-j(1/2-3/q)}\,\|f\|_q\,\|g\|_B \tag{15.27}$$

where $\|g\|_B$ denotes the norm of g in the homogeneous Besov space $\dot{B}_2^{1/2,\infty}$.

This estimates implies $\sum_{-\infty}^{\infty} A_j \in \dot{B}_2^{1/2-3/q,\infty}$ and the hypothesis $q \in (3/2, 6)$ is not needed.

We now turn to the second series $B(x) = \sum_{-\infty}^{\infty} B_j(x)$. Here $q < 6$ is needed. This implies that two exponents $s > 3$ and $r \geq q$ exist with $\frac{1}{s} + \frac{1}{r} = \frac{1}{2}$. We then have

$$\|\Delta_j(g)\|_s \leq C\,2^{j(1-3/s)}\,\|g\|_B \tag{15.28}$$

which implies

$$\|S_j(g)\|_s \leq C'\,2^{j(1-3/s)}\,\|g\|_B\,. \tag{15.29}$$

Bernstein's inequalities yield

$$\|\Delta_j(f)\|_r \leq C\,2^{3j(1/q-1/r)}\,\|f\|_q\,. \tag{15.30}$$

All together these estimates give

$$\big\|S_{j-2}(g)\,\Delta_j(f)\big\|_2 \leq C\,2^{-j(1/2-3/q)}\,\|f\|_q\,\|g\|_B \tag{15.31}$$

as expected.

The last series in the paraproduct algorithm are $\sum_{-\infty}^{\infty} \Delta_j(f)\,\Delta_{j'}(f)$ where $|j'-j| \leq 2$. We consider $\sum_{-\infty}^{\infty} \Delta_j(f)\,\Delta_j(g)$ since the four other series can receive a similar treatment. Since we cannot expect any cancellation in $\Delta_j(f)\,\Delta_j(g)$, lemma 18 should be used with $s > 0$.

We separately treat the two cases $3/2 < q \leq 2$ and $2 \leq q < 6$.

In the first case, the L^1 norm of $C_j = \Delta_j(f)\,\Delta_j(g)$ is estimated by Hlder's inequality and one obtains $\|C_j\|_1 \le C\,2^{j(3/q-2)}$. Then $\sum_{-\infty}^{\infty} C_j(x)$ belongs to $\dot{B}_2^{1/2-3/q,\infty}$. Since $\dot{B}_1^{2-3/q,\infty}$ is contained in $\dot{B}_2^{1/2-3/q,\infty}$, this ends the proof.

When $2 \le q < 6$, an exponent r is defined by $\frac{1}{2} + \frac{1}{q} = \frac{1}{r}$ and Hlder's inequality yields

$$\|\Delta_j(f)\,\Delta_j(g)\|_r \le C\,2^{-j/2}\,\|f\|_q\,\|g\|_B \tag{15.32}$$

which implies $\sum_{-\infty}^{\infty} C_j \in \dot{B}_r^{1/2,\infty}$. Here again $\dot{B}_r^{1/2,\infty}$ is contained in $\dot{B}_2^{1/2-3/q,\infty}$ and the proof is completed.

It should be observed that theorem 15.2 fails if $q = 3/2$ or $q = 6$.

In both cases, we start a simple observation. The function $g(x) = \frac{1}{|x|}$ belongs to $\dot{B}_2^{1/2,\infty}$. Indeed this function belongs to $\dot{B}_q^{-(1-1/3),\infty}$ for $q \ge 1$. We then study the linear operator defined by the pointwise multiplication by $\frac{1}{|x|}$.

For studying this operator, the following lemma will be used.

Lemma 19. *There exists a constant C_0 such that*

$$\|f\|_2 \le C_0 \sup\left\{\|e^{i\omega\cdot x} f(x)\|_{\dot{B}_2^{0,\infty}}\,;\ \omega \in \mathbb{R}^3\right\} \tag{15.33}$$

for $f \in L^2(\mathbb{R}^3)$.

The proof of (15.33) is almost obvious. If $\hat{f}$ is compactly supported and vanishes when $|\xi| > R$, then we have

$$c_1\|f\|_2 \le \left\|e^{i\omega\cdot x} f(x)\right\|_{\dot{B}_2^{0,\infty}} \le c_2\|f\|_2 \tag{15.34}$$

whenever $2R = |\omega|$. Here c_2 and c_1 are two positive constants. Then (15.34) implies (15.33). When $\hat{f}$ is no longer compactly supported, it suffices to approach f by such functions.

Let us assume for a while that the pointwise product between a function g in $L^6(\mathbb{R}^3)$ with $\frac{1}{|x|}$ belongs to $\dot{B}_2^{0,\infty}$. Since $g(x)$ and $e^{i\omega\cdot x}\,g(x)$ have the same L^6 norms, this would imply the following

$$\left\|f(x)\,|x|^{-1}\right\|_2 \le C\|f\|_6\,. \tag{15.35}$$

This is obviously wrong since $|x|^{-1}$ does not belong to $L^3(\mathbb{R}^3)$.

The other limiting case is $q = 3/2$.

Let us denote by $\theta(x)$ a cut-off function such that $\theta(x) = 1$ on $|x| \leq 1/4$ and $\theta(x) = 0$ if $|x| \geq 1/2$. We then consider $f(x) = |x|^{-2} \,|\log |x||^{-1} \theta(x)$ which belongs to $L^{3/2}$. The pointwise product between $f(x)$ and $\frac{1}{|x|}$ is $|x|^{-3} \,|\log |x||^{-1} \theta(x)$. This product is a pointwise function which is not locally integrable at 0. Therefore it cannot belong to $\dot{B}_2^{-3/2,\infty}$ and it is not a tempered distribution.

16 T. Kato's algorithm : an abstract lemma

The goal of this algorithm is to construct mild solutions $u \in C([0,\infty); E)$ to the Navier-Stokes equations when the functional Banach space E is adapted to the Navier-Stokes equations.

More precisely we would like to prove the following conjecture :

(16.1) there exists a positive number α and a constant C
such that for $u_0 \in E$ satisfying $\|u_0\|_E < \alpha$ and $\operatorname{div} u_0 = 0$,
there exists a mild solution $u \in C([0,\infty); E)$
to the Navier-Stokes equations such that $u(\cdot, 0) = u_0$ and

$$\sup_{t\geq 0} \|u(\cdot,t)\|_E \leq C\|u_0\|_E \,. \tag{16.2}$$

As it was earlier mentioned, the condition $\|u_0\|_E < \alpha$ cannot be given a meaning if the functional space E is not adapted to the Navier-Stokes equations. Indeed the Navier-Stokes equations are invariant under a certain action of the affine group and the sufficient condition $\|u_0\|_E < \alpha$ should reflect this invariance.

T. Kato's algorithm relies on the following abstract lemma.

Lemma 20. *Let Y be a Banach space and let $B : Y \times Y \to Y$ be a continuous bilinear operator : there exists a constant C_0 such that*

$$\|B(y,z)\| \leq C_0 \,\|y\| \,\|z\| \tag{16.3}$$

for $y, z \in Y$, $\|\cdot\|$ denoting the norm in Y.

Then for $a \in Y$ such that $\|a\| < \frac{1}{4C_0}$, there exists at least a solution $x \in Y$ to the equation

$$x = a + B(x,x)\,. \tag{16.4}$$

For proving this lemma we inductively define a sequence x_k, $k \in \mathbb{N}$ (and $k = 0$) by $x_0 = 0$ and $x_{k+1} = a + B(x_k, x_k)$. We then obtain

(16.5)
$$x_{k+1} - x_k = B\big(x_k - x_{k-1}\,,\, x_k\big) + B\big(x_{k-1}\,,\, x_k - x_{k-1}\big)$$

which implies

(16.6)
$$\big\|x_{k+1} - x_k\big\| \le C_0 \big\|x_k - x_{k-1}\big\| \big(\|x_{k-1}\| + \|x_k\|\big)\,.$$

For estimating $\|x_{k+1} - x_k\|$, we construct an increasing sequence r_k of real numbers by $r_0 = 0$, $r_1 = \|a\|$ and $r_{k+1} = \|a\| + r_k^2$. We then have

(16.7)
$$r_{k+1} - r_k = C_0\big(r_k - r_{k-1}\big)\big(r_k + r_{k-1}\big)$$

and r_k is actually increasing. This sequence r_k converges to r if and only if $4C_0\|a\| \le 1$. This limit r is the smallest solution of the scalar equation $t = \|a\| + C_0 t^2$. This equation is a special case of (16.4) where $Y = \mathbb{R}$ and $B(t, t) = C_0 t^2$. An obvious induction yields $\|x_{k+1} - x_k\| \le r_{k+1} - r_k$. Therefore $\sum_0^\infty \|x_{k+1} - x_k\| \le r$ and x_k, $k \in \mathbb{N}$ is a Cauchy sequence which tends to a limit x. We have $\|x\| \le r = \frac{1 - \sqrt{1 - 4\|a\| C_0}}{2C_0} \le 2\|a\|$.

This well-known proof is interesting since it is sharp. Indeed $\|a\| < \frac{1}{4C_0}$ cannot be replaced by $\|a\| < \frac{1+\varepsilon}{4C_0}$ where ε is positive. A counter-example is precisely given by the scalar equation $t = \|a\| + C_0 t^2$.

Returning to lemma 20, it is completed by the following remark.

Lemma 21. *We keep the same assumptions and notations as in lemma 20. Then the equation* (16.4) *has a unique solution x such that* $\|x\| < \frac{1}{2C_0}$.

We first observe that the solution given by the iterative scheme satisfies $\|x\| \le 2\|a\| < \frac{1}{2C_0}$.

If y is another such solution and $y = x + z$, we then have $z = B(x, z) + B(z, y)$ which implies $\|z\| \le C_0 \|x\|\,\|z\| + C_0 \|y\|\,\|z\|$. If $z \ne 0$, this reads $1 \le C_0\big(\|x\| + \|y\|\big) < 1$.

17 Kato's algorithm applied to Navier-Stokes equations : a straightforward example

We want to apply lemma 20 to Navier-Stokes equations. More precisely we want Y to be the Banach space $C([0, \infty); E)$ where E is a functional space which is adapted to the Navier-Stokes equations. We want a to be $S(t)u_0 \in C([0,\infty); E)$. Finally the Navier-Stokes equations read

$$(17.1) \qquad u = u(\cdot,t) = S(t)u_0 + B(u,u) = a + B(u,u)$$

where

$$(17.2) \qquad B(u,v) = \mathbb{P}\int_0^t S(t-s)\,\partial_j(u_j v)(s)\,ds\,.$$

Some explanations may be useful. Here and in what follows, $a_j b_j$ means $a_1b_1 + a_2b_2 + a_3b_3$. We will write $u \in C([0,\infty); E)$ instead of the unpleasant $u \in C([0,\infty); E^3)$. The Banach space $Y = C([0,\infty); E)$ will be given the norm

$$(17.3) \qquad \sup_{t\geq 0} \|u(\cdot,t)\|_E\,.$$

More important are the following remarks about the bilinear operator B. The Banach spaces E which will be used are invariant under the Riesz transformations. We can therefore forget the Leray projector $\mathbb{P}$ which acts boundedly on E.

In a more sophisticated version of Kato's algorithm (section 19) an L^∞ estimate will also be needed. Since the Leray projector $\mathbb{P}$ is not bounded on $L^\infty(\mathbb{R}^n)$, it should be incorporated inside the integral sign and the three operators $\mathbb{P}$, $S(t-s)$ and ∂_j should be glued together. The resulting operator $\Gamma_j(t-s)$, $j = 1, 2$ or 3, is a convolution with $(t-s)^{-2}\,\omega_j\big(x(t-s)^{-1/2}\big)$, $j = 1,2$ or 3. Elementary calculations show that $\omega_j \in C^\infty(\mathbb{R}^3)$ with $\omega_j(x) = 0(|x|^{-4})$ at infinity. Therefore ω_j belongs to $L^1(\mathbb{R}^3)$ and the operator norm of

$$(17.4) \qquad \mathbb{P}\,S(t-s)\partial_j \;:\; L^\infty(\mathbb{R}^3) \to L^\infty(\mathbb{R}^3)$$

is $C(t-s)^{-1/2}$ where C is an absolute constant. This remark applies as well if $L^\infty(\mathbb{R}^3)$ is replaced by any functional Banach space X whose norm is translation invariant.

In (17.192) u and v are vector fields satisfying

$$\partial_1 u_1 + \partial_2 u_2 + \partial_3 u_3 = \partial_1 v_1 + \partial_2 v_2 + \partial_3 v_3 = 0\,.$$

However it can be proved that for most of the estimates we have in mind, this fact is useless [57]. That is why at the proof level the true bilinear operator $B(u,v)$ will be replaced by their scalar counterparts defined as

(17.5)

$$B_j(f,g) = \int_0^t S(t-s)\,\partial_j(fg)\,(s)\,ds\,, \qquad 1 \leq j \leq 3\,,$$

where $f = f(x,t)$ and $g = g(x,t)$ both belong to $C([0,\infty); E)$.

We can even move further and consider

$$B(f,g) = \int_0^t S(t-s)\,\Lambda(fg)\,(s)\,ds \tag{17.6}$$

where $\Lambda = \sqrt{-\Delta}$ is the celebrated Calderón's operator. Then $B_j = -iR_jB$ where R_j denotes the Riesz transformation. Since E is assumed to be invariant under the action of R_j, any estimate for $B(f,g)$ will imply the corresponding estimate for $B_j(f,g)$.

In our first example of Kato's algorithm, the functional Banach space E is defined by a pointwise estimate on the Fourier transform $\hat{f}$ of f. It reads

$$|\hat{f}(\xi)| \le C|\xi|^{-2}. \tag{17.7}$$

In other words $\Delta f \in PM(\mathbb{R}^3)$ where following Kahane's notations, a distribution S is a pseudo-measure if and only if its Fourier transform $\hat{S}$ belongs to $L^\infty(\mathbb{R}^3)$.

Therefore E is not a separable Banach space. Instead it is a dual space F^* where $g \in F$ if and only if $g = \Delta h$ and h belongs to the Wiener algebra [41]. Therefore $u \in C([0,\infty); E)$ means the following two properties

(17.8) u is continuous from $(0,\infty)$ to E when E is given its norm topology

(17.9) $u(\cdot,t) \rightharpoonup u(\cdot,0)$ with respect to $\sigma(E,F)$ as t tends to 0.

We then write $Y = C([0,\infty); E)$ and have

Lemma 22. *The bilinear operator B defined by (17.6) is continuous from $Y \times Y$ into Y.*

We will only prove the bilinear estimate

$$\|B(f,g)\|_Y \le C\,\|f\|_Y\,\|g\|_Y \tag{17.10}$$

when

$$\|f\|_Y = \sup_{t\ge 0} \|f(\cdot,t)\|_E\,. \tag{17.11}$$

The proof of the required continuity with respect to the time variable is identical to the one which will be given when $E = L^{3,\infty}$. For that reason, this part will be omitted.

Returning to (17.200), we observe that $|\hat{f}(\xi)| \leq C|\xi|^{-2}$ and $|\hat{g}(\xi)| \leq C|\xi|^{-2}$ imply $|(fg)\hat{}(\xi)| \leq C|\xi|^{-1}$. On the Fourier transform side,

$$[B(f,g)]\hat{}(\xi) = \int_0^t |\xi|\, e^{-(t-s)|\xi|^2}\, (fg)\hat{}(\xi)\, ds$$

which implies

$$|[B(f,g)]\hat{}(\xi)| \leq C \int_0^t |\xi|\, e^{-(t-s)|\xi|^2} |\xi|^{-1}\, ds \leq C|\xi|^{-2}.$$

We have proved the following result

Theorem 17.1. *If the adapted Banach space E is defined by (17.197), there exists a positive number η such that for every initial condition satisfying $\|u_0\|_E < \eta$, there exists a mild solution*

$$u \in C([0,\infty); E)$$

to the Navier-Stokes equations such that $u(x,0) = u_0(x)$ and

$$\sup_{t\geq 0} \|u(x,t)\|_E \leq 2\|u_0\|_E. \tag{17.12}$$

Moreover this solution is uniquely defined by the weaker condition

$$\sup_{t\geq 0} \|u(x,t)\|_E < 2\eta. \tag{17.13}$$

18 Kato's algorithm applied to the Navier-Stokes equations : the Lorentz space $L^{3,\infty}(\mathbb{R}^3)$

Let us remind the reader with the general definition of the Lorentz spaces $L^{p,q}(\mathbb{R}^n)$. If f is a measurable function defined on $\mathbb{R}^n$, we let E_j, $j \in \mathbf{Z}$, be the set of points $x \in \mathbb{R}^n$ for which $2^j \leq |f(x)| < 2^{j+1}$. If $|E_j|$ denotes the Lebesgue measure of E_j, then $f \in L^{p,q}(\mathbb{R}^n)$ means

$$|E_j|^{1/p}\, 2^j \in \ell^q(\mathbf{Z}). \tag{18.1}$$

Let us observe some formal similarities between this definition and the definition of the Besov spaces. Writing $f_j = \mathbf{1}_{E_j} f$ where $\mathbf{1}_E$ is the indicator function of E, (18.1) reads

$$\|f_j\|_p \in \ell^q(\mathbf{Z}) \tag{18.2}$$

and we have $f = \sum_{-\infty}^{\infty} f_j$.

But in contrast with what is happening in the Besov space case, the mapping $f \to (f_j)_{j\in\mathbf{Z}}$ is not a linear one.

We now assume $1 < p < \infty$ and $q = \infty$. Then (18.1) can be rewritten as

$$\left|\{x\,;\,|f(x)| > \lambda\}\right| \leq C\lambda^{-p}\,. \tag{18.3}$$

If E is any measurable set with a finite measure, (18.3) implies

$$\int_E |f(x)|\,dx \leq C'|E|^{1/p'} \tag{18.4}$$

where $\frac{1}{p} + \frac{1}{p'} = 1$.

This fails when $p = 1$. Obviously (18.4) implies $f \in L^{p,\infty}(\mathbb{R}^n)$.

Then (18.4) can easily be used for defining a norm on $L^{p,\infty}$. It suffices to write

$$\|f\|_{(p,\infty)} = \sup\left\{|E|^{-1/p'}\int_E |f(x)|\,dx\,;\,E \in \mathcal{B}\right\} \tag{18.5}$$

where $\mathcal{B}$ is the collection of all Borel sets with a finite and positive measure.

An other access to the norm is given by the following observation. A function $a(x)$ is an atom if $a(x) = |E|^{-1/p'}\,\mathbf{1}_E(x)$ for some E in $\mathcal{B}$. We then consider the Banach space consisting of all $f(x) \in L^{p'}(\mathbb{R}^n)$ which admit a decomposition

$$f(x) = \sum_0^\infty \lambda_j\, a_j(x) \tag{18.6}$$

where $a_j(x)$ is a sequence of atoms and where $\sum_0^\infty |\lambda_j| < \infty$.

The norm of f being the obvious quotient norm given by the infimum of $\sum_0^\infty |\lambda_j|$ computed over all possible expansions of f.

It is not difficult to check that this Banach space is identical to $L^{p',1}(\mathbb{R}^n)$. Then $L^{p,\infty} = E$ is the dual space of $L^{p',1} = F$. This Banach space F is separable while E is not.

We now return to $L^{3,\infty}(\mathbb{R}^3)$. If $f \in L^{3,\infty}(\mathbb{R}^3)$, we have

$$\int_E |f(x)|^2\,dx \le C|E|^{1/3}\,, \qquad E \in \mathcal{B} \tag{18.7}$$

and it is interesting to compare this property to the definition of the Morrey-Campanato space where E is a ball.

As it was stated many times, $u(x,t) \in C([0,\infty);E)$, $E = L^{3,\infty}(\mathbb{R}^3)$ means $u(x,t) \rightharpoonup u(x,0)$ in $\sigma(E,F)$ as t tends to 0 together with

$$\lim_{t\to t_0} \left\| u(\cdot,t) - u(\cdot,t_0) \right\|_E = 0\,, \qquad t_0 > 0\,. \tag{18.8}$$

We then have

Lemma 23. *If $E = L^{3,\infty}(\mathbb{R}^3)$ and $Y = C([0,\infty);E)$ then the bilinear operator $\mathcal{B}$ is continuous from $Y \times Y$ into Y.*

Let us begin with a general fact.

We start with a kernel $K_t(x,y)$, $t > 0$, $x \in \mathbb{R}^3$, $y \in \mathbb{R}^3$, fulfilling the following estimate

$$|K_t(x,y)| \le Ct^{-3}\big(1 + |x-y|/t\big)^{-4} \tag{18.9}$$

and denote by $P_t : L^p(\mathbb{R}^3) \to L^p(\mathbb{R}^3)$, $1 \le p \le \infty$, the operator defined by

$$P_t f(x) = \int K_t(x,y)\, f(y)\, dy\,. \tag{18.10}$$

We then have a pointwise inequality

$$|P_t f|(x) \le C\big(\varphi_t * |f|\big)(x) \tag{18.11}$$

where $\varphi(x) = (1+|x|)^{-4}$. Therefore

$$\|P_t f\|_E \le C\|f\|_E \tag{18.12}$$

whenever the functional Banach space E is translation invariant and is a lattice. That means that $f_2 \in E$ and $|f_1(x)| \le |f_2(x)|$ implies $f_1 \in E$. This remark and (18.215) apply to $L^{p,\infty}(\mathbb{R}^3)$ when $1 < p < \infty$.

Next we consider a real number $\alpha \in (0,3)$ and an exponent p belonging to $(1, 3/\alpha)$. We then study a linear operator T which maps continuous functions $f \in C([0,\infty); E)$, $E = L^{p,\infty}$, into functions of $x \in \mathbb{R}^3$ by

$$g(x) = \int_0^\infty P_t f(\cdot, t)\, t^{\alpha-1}\, dt\,. \tag{18.13}$$

Keeping these notations we have

Theorem 18.1. *If $\alpha \in (0,3)$ and $p \in (1, 3/\alpha)$, then*

$$\|g\|_{(q,\infty)} \le C \sup_{t\ge 0} \|f(\cdot,t)\|_{(p,\infty)} \tag{18.14}$$

where $q = \frac{3p}{3-\alpha p}$ and $\|\cdot\|_{(p,\infty)}$ is the weak-L^p norm.

The proof is trivial but will be given for the reader's convenience. Let $\lambda \in (0,\infty)$ be a threshold. We consider the set E of points x for which $|g(x)| > \lambda$ and we want to estimate the Lebesgue measure $|E|$ of E. We then split $\int_0^\infty P_t f(\cdot,t) t^{\alpha-1}\, dt$ into $\int_0^\tau + \int_\tau^\infty = u(x) + v(x)$. The relation between τ and λ will soon be clarified.

We first estimate $\|v\|_\infty$. Indeed as a function of y, $K_t(x,\cdot)$ belongs to $L^{p',1}$ where $1/p' + 1/p = 1$. Moreover $\|K_t(x,\cdot)\|_{p',1} \le Ct^{-3/p}$ and this implies

$$\|P_t f(\cdot,t)\|_\infty \le Ct^{-3/p} \sup_{t\ge 0} \|f(\cdot,t)\|_{(p,\infty)}\,. \tag{18.15}$$

Therefore $\|v\|_\infty \le C\int_\tau^\infty t^{\alpha-3/p-1}\, dt = C'\tau^{\alpha-3/p}$. We now define τ by $C'\tau^{\alpha-3p} = \lambda/2$. If $|g(x)| > \lambda$, that forces $|u(x)| > \lambda/2$. But

$$\|u\|_{(p,\infty)} \le C \int_0^\tau \|f(\cdot,t)\|_{(p,\infty)}\, t^{\alpha-1}\, dt = C'\tau^\alpha$$

(here $p > 1$ is crucially needed). Finally

$$\big|\{x\,;\, |u(x)| > \lambda/2\}\big| \le C''\Big(\frac{\lambda}{\tau^\alpha}\Big)^{-p} = C_0 \lambda^{-q}$$

where $q = \frac{3p}{3-\alpha p}$ as announced.

We now treat a very specific example. We consider

$$g(x,t) = \int_0^t S(t-s)\, \Lambda\, f(\cdot,s)\, ds$$

where $\sup_{s\geq 0}\|f(\cdot,s)\|_{(3/2,\infty)} \leq 1$ and want to prove that a constant C exists for which

$$\sup_{t\geq 0}\|g(\cdot,t)\|_{(3,\infty)} \leq C .$$

For reducing this example to the framework given in theorem 18.1, we first change s into $t-s$ and write

$$g(x,t) = \int_0^\infty S(s)\,\Lambda\,F(\cdot,s)\,ds$$

with $F(\cdot,s) = f(\cdot,t-s)$ if $0\leq s\leq t$ and $F(\cdot,s)=0$ if $s>t$.

Finally $\Lambda\,S(s)$ is a convolution operator. The corresponding kernel $K_s(x,y)$ is $s^{-2}\,\omega\big(\frac{x-y}{\sqrt{s}}\big)$ where $|\omega(x)| \leq C(1+|x|)^{-4}$. This leads to a final change of variables $s=\tau^2$ and theorem 18.1 applies with $\alpha=1$.

We now investigate continuity with respect to the t-variable.

Lemma 24. *Let us assume that $f(\cdot,s)\in C([0,\infty);E)$, $E=L^{3/2,\infty}$ where this continuity with respect to the t variable is defined by* (17.198) *and* (17.199).

Then $g(\cdot,t)\in C([0,\infty);F)$ where $F=L^{3,\infty}$ and the continuity is similarly defined.

We begin with the required weak continuity at $t_0=0$. Our first observation is a trivial one. Indeed

$$\begin{aligned}\|g(\cdot,t)\|_{3/2,\infty} &\leq C\int_0^t (t-s)^{-1/2}\,\|f(\cdot,s)\|_{(3/2,\infty)}\,ds\\ &\leq 2Ct^{1/2}\sup_{0\leq s\leq t}\|f(\cdot,s)\|_{(3/2,\infty)}\,.\end{aligned}$$

Therefore $\lim_{t\downarrow 0}\|g(\cdot,t)\|_{3/2,\infty}=0$ which suffices to imply the required weak continuity.

For proving the required continuity at $t_0>0$, we compute $\|g(\cdot,t')-g(\cdot,t)\|_E$ when $0<a<t<t'<2a$. This positive a exists if $|t-t_0|<t_0/3$ and $|t'-t_0|<t_0/3$.

For symplifying the notations, we assume

$$\|f(\cdot,t)\|_E \leq 1 \qquad \text{for} \quad t\geq 0\,.$$

We then split $f(\cdot,t)$ into the sum $f_1(\cdot,t)+f_2(\cdot,t)$ where $f_1(\cdot,t)=0$ for $t\geq a/2$, $f_2(\cdot,t)=0$ for $t\leq a/4$, $f_2(\cdot,t)$ is continuous from $[0,\infty)$ into

E when E is given its strong norm topology and finally $\|f_1(\cdot,t)\|_E \le 1$, $\|f_2(\cdot,t)\| \le 1$ for $t \ge 0$. Then

$$\begin{aligned} g(\cdot,t) &= \int_0^t S(t-s)\,\Lambda\, f(\cdot,s)\,ds \\ &= \int_0^t S(t-s)\,\Lambda\, f_1(\cdot,s)\,ds + \int_0^t S(t-s)\,\Lambda\, f_2(\cdot,s)\,ds \\ &= u(\cdot,t)+v(\cdot,t)\,. \end{aligned}$$

We first treat $u(\cdot,t)=\int_0^t S(t-s)\,\Lambda f_1(\cdot,s)\,ds$. We then have

$$\begin{aligned} \left\|\frac{\partial}{\partial t}u(\cdot,t)\right\|_{(3,\infty)} &\le \int_0^\infty \left\|\Delta^2 S(t-s)\,(\Lambda^{-1}f_1)(\cdot,s)\right\|_{(3,\infty)} \\ &\le C\int_0^{a/2}(t-s)^{-2}\,ds\,. \end{aligned}$$

This last estimate is coming from the boundedness of $\Lambda^{-1}: L^{3/2,\infty}\to L^{3,\infty}$. Concerning v, we write

$$\begin{aligned} v(\cdot,t')-v(\cdot,t) &= \int_0^{t'} S(\tau)\,\Lambda f_2(\cdot,t'-\tau)d\tau - \int_0^t S(\tau)\,\Lambda f_2(\cdot,t-\tau)d\tau \\ &= \int_t^{t'} S(\tau)\,\Lambda\, f_2(\cdot,t'-\tau)\,d\tau\, + \\ &\quad + \int_0^t S(\tau)\,\Lambda[f_2(\cdot,t'-\tau)-f_2(\cdot,t-\tau)]\,d\tau = U+V\,. \end{aligned}$$

Concerning U, we once move factor out the Calderón operator which reads

$$U = \int_t^{t'} [S(\tau)\Delta]\,\Lambda^{-1}\, f_2(\cdot,t'-\tau)\,d\tau$$

and implies

$$\|U\|_{3,\infty} \le \int_t^{t'} \left\|Q(\tau)\,\Lambda^{-1}f_2(\cdot,t'-\tau)\right\|_{3,\infty}\frac{d\tau}{\tau}\,.$$

The operator $Q(\tau)=\tau\,\Delta\,\exp(\tau\Delta)$, $\tau\ge 0$, is uniformly bounded on $L^{3,\infty}$. We finally obtain

$$\|U\|_{3,\infty} \le C\,\log\frac{t'}{t}\,.$$

Concerning V, it suffices to use theorem 18.1 together with the continuity with respect to the t variable of f_2. We then have $\|f_2(\cdot,t'-\tau)-f_2(\cdot,t-$

$\tau)\|_{3/2,\infty} \leq \varepsilon$ if $|t'-t| \leq \eta$ and $0 \leq \tau \leq t \leq t' \leq 2a$. It implies $\|V\|_{3,\infty} \leq C\varepsilon$ as expected.

We proved the following statement.

Theorem 18.2. *There exists a positive number η such that for every $u_0(x)$ in $L^{3,\infty}(\mathbb{R}^3)$ fulfilling*

$$\|u_0\|_{3,\infty} < \eta \quad \text{and} \quad \operatorname{div} u_0(x) = 0 \tag{18.16}$$

there is a solution $u(x,t)$ of the Navier-Stokes equations with the following properties

$$u \in C([0,\infty);\, L^{3,\infty}(\mathbb{R}^3)) \tag{18.17}$$

$$u(x,0) = u_0(x) \tag{18.18}$$

$$\sup_{t\geq 0} \|u(x,t)\|_{(3,\infty)} \leq 2\|u_0\|_{(3,\infty)}\,. \tag{18.19}$$

Moreover this solution is uniquely defined by (18.220), (18.221) *and*

$$\sup_{t\geq 0} \|u(x,t)\|_{(3,\infty)} < 2\eta\,. \tag{18.20}$$

19 Kato's algorithm : the Lebesgue space $L^3(\mathbb{R}^3)$

Theorem 18.2 is surprising since the continuity of $B : Y \times Y \to Y$ is failing in a more natural setting given by $Y = C([0,\infty); L^3(\mathbb{R}^3))$.

For treating this example, T. Kato made the following crucial observations. First the linear evolution $S(t)u_0$ belongs to a much narrower space $Z \subset Y$ and secondly the iterative scheme which is being used for solving (17.191) is indeed confined inside Z.

The definition of Z heavily depends on the adapted functional space E we are working with.

When $E = L^3(\mathbb{R}^3)$ or the Morrey-Campanato space M^2, the definition of Z relies on the L^∞-norm of $u(\cdot,t)$. In other cases, this definition is much more involved and will be unveiled in section 22.

We begin with a simple observation which was proved in section 8 (lemma 9).

Lemma 25. *If the functional Banach space E is adapted to the Navier-Stokes equations, there exists a constant C such that for $f \in E$ and $t > 0$, we have*

$$\|S(t)f\|_\infty \leq Ct^{-1/2}\|f\|_E\,. \tag{19.1}$$

This estimate leads to the following definition of Z as proposed by T. Kato : Z is the subspace of $Y = C([0,\infty); E)$ consisting of all $u(x,t) \in Y$ such that

$$\sup_{t\geq 0} t^{1/2} \|u(\cdot,t)\|_\infty = \gamma < \infty \tag{19.2}$$

and $\|u\|_Z = \gamma + \sup_{t\geq 0} \|u(\cdot,t)\|_E$. Lemma 25 implies the following

$$\|S(t)u_0\|_Z \leq C_1\|u_0\|_E \tag{19.3}$$

and Kato's program will depend on the following crucial estimate

$$\|B(u,v)\|_Z \leq C_0 \|u\|_Z \|v\|_Z\,. \tag{19.4}$$

We illustrate this approach when $E = L^3(\mathbb{R}^3)$ and prove (19.4). We begin with the first half of the Z-norm and prove

$$\sup_{t\geq 0} \|B(u,v)\|_3 \leq C \|u\|_Z \|v\|_Z\,. \tag{19.5}$$

Indeed $B(u,v) = w(\cdot,t) = \int_0^t \mathbb{P}\, S(t-s)\, \partial_j(u_j v)\, ds$ and $\mathbb{P}\, S(t-s)\partial_j$ is a convolution operator with a (matrix valued) function $(t-s)^{-2}\,\omega((t-s)^{-1/2}x)$. As it was observed, the L^1-norm of this function is $(t-s)^{-1/2}\,\|\omega\|_1$. Therefore

$$\|w(\cdot,t)\|_3 \leq \|\omega\|_1 \int_0^t (t-s)^{-1/2}\,\|uv\|_3\, ds\,.$$

We then observe that $\|uv\|_3 \leq \|u\|_\infty\|v\|_3 \leq s^{-1/2}\,\|u\|_Z\,\|v\|_Z$ and the computation ends with the following trivial remark

$$\int_0^t (t-s)^{-1/2}\, s^{-1/2}\, ds = \pi\,. \tag{19.6}$$

This simple approach fails if one tries to estimate the L^∞ norm. Indeed $\|u\|_\infty \leq s^{-1/2}\,\|u\|_Z$, $\|v\|_\infty \leq s^{-1/2}\,\|v\|_Z$ lead to a divergent integral.

But this problem can be easily fixed. Indeed one uses the duality between L^4 and $L^{4/3}$ which yields

$$\|w(\cdot,t)\|_\infty \leq C \int_0^t (t-s)^{-7/8}\,\|uv\|_4\, ds$$

since the $L^{4/3}$ norm of $(t-s)^{-2}\,\omega((x(t-s)^{-1/2})$ is $C(t-s)^{-7/8}$. Next $\|uv\|_4 \leq \|u\|_8\,\|v\|_8$ and these L^8 norms are trivialy estimated. Indeed $\|u\|_3 \leq \|u\|_Z$ while $\|u\|_\infty \leq s^{-1/2}\|u\|_Z$ which yields $\|u\|_8 \leq s^{-5/16}\,\|u\|_Z$.

We now need to show that $B(u,v)$ belongs to $C([0,\infty);L^3)$ if u and v belong to Z. The main issue is the continuity at 0. For proving the fundamental property

$$\lim_{t\downarrow 0} \|B(u,v)\|_3 = 0\,, \qquad u \in Z\,,\ v \in Z \tag{19.7}$$

we need to start with a simple observation. If $u_0 \in L^3(\mathbb{R}^3)$, we have

$$\lim_{t\downarrow 0} t^{1/2}\, \|S(t)u_0\|_\infty = 0\,. \tag{19.8}$$

This allows us to modify and sharpen the definition of Z. We impose for $u \in Z$

$$\lim_{t\downarrow 0} t^{1/2}\, \|u(\cdot,t)\|_\infty = 0\,. \tag{19.9}$$

The continuity of the bilinear operator $B : Z \times Z \to Z$ is proved with the same argument we previously used and (19.230) follows from the definition of Z.

We can state

Theorem 19.1. *There exists a positive constant η such that for any initial condition $u_0(x) \in L^3(\mathbb{R}^3)$ such that*

$$\|u_0\|_3 < \eta \quad \textit{and} \quad \operatorname{div} u_0(x) = 0 \tag{19.10}$$

there exists a mild solution

$$u(x,t) \in C\big([0,\infty); L^3(\mathbb{R}^3)\big) \tag{19.11}$$

to the Navier-Stokes equations such that $u(x,0) = u_0(x)$.

This solution satisfies

$$\sup_{t\geq 0} \big(\|u(\cdot,t)\|_3 + t^{1/2}\, \|u(\cdot,t)\|_\infty\big) \leq C\|u_0\|_3 \tag{19.12}$$

where C is an absolute constant.

Finally there exists a second positive constant $\beta < 2\eta$ such that this solution is uniquely defined by

$$\sup_{t>0} \|u(\cdot,t)\|_3 + t^{1/2}\, \|u(\cdot,t)\|_\infty < \beta\,. \tag{19.13}$$

Let us observe that Kato's algorithm is here applied to the Z-norm and that β is no longer 2η since we need to take in account (19.3).

Before leaving this example, let us make a last remark. We already know that our solution $u(x,t)$ satisfies

$$\lim_{t\to 0} t^{1/2}\,\|u(\cdot,t)\|_\infty = 0\,. \tag{19.14}$$

Indeed uniqueness can be proved with (19.14) instead of (19.13). Let us sketch the proof of this observation.

Instead of studying solutions u in $C([0,\infty);E)$ we could as well consider solutions in $C([0,T];E)$ where T is a positive constant. All the estimates we proved in the first setting can be proved in this second framework and the constants remain unchanged : they do not depend on T.

If (19.14) is satisfied, we can chose T small enough in such a way that

$$\sup_{0\le t\le T}\left(\|u(\cdot,t)\|_3 + t^{1/2}\,\|u(\cdot,t)\|_\infty\right) < \beta\,.$$

This implies uniqueness on $[0,T]$.

There is a much deeper statement which will be unveiled in the next section.

20 Uniqueness of L^3 valued mild solutions

In this section a simplified proof of Lemarié-Rieusset's theorem will be given. The original and more involved proof will be described in section 26.

Theorem 20.1. *Let $u(x,t)\in C([0,T];L^3(\mathbb{R}^3))$ and $v(x,t)\in C([0,T];L^3(\mathbb{R}^3))$ be two mild solutions of Navier-Stokes equations such that $u(x,0)=v(x,0)=u_0(x)$.*

Then $u=v$ on $[0,T]$.

Let us denote by τ the supremum of the set of $\sigma\in[0,T]$ such that $u=v$ on $[0,\sigma]$. By continuity, $u=v$ on $[0,\tau]$. If $\tau<T$ we will prove that $u=v$ on $[0\,,\,\tau+\varepsilon]$ for some positive ε. This forces $\tau=T$ as announced.

We now consider $\overline{u}(\cdot,t)=u(\cdot\,,\,t+\tau)$ and $\overline{v}(\cdot,t)=v(\cdot\,,\,t+\tau)$. These two functions are defined on $[0\,,\,T-\tau]$. Both are mild solutions to the Navier-Stokes equations (13.2) and they coincide when $t=0$.

Changing the notations, we conclude that theorem 20.1 will be proved as soons as we can prove the following statement

$$u = v \quad \text{on} \quad [0, \varepsilon] \tag{20.1}$$

for some positive ε.

For proving (20.238) one writes

$$u = S(t)u_0 + B(u,u) \quad , \quad v = S(t)u_0 + B(v,v) \tag{20.2}$$

and $v = u + w$. Then

$$w = B(u+w, u+w) - B(u,u) = B(v,w) + B(w,u).$$

Our goal is to estimate

$$\eta(t) = \sup_{0 \le s \le t} \|w(\cdot, s)\|_{(3,\infty)} \tag{20.3}$$

and prove

$$\eta(t) \le \frac{1}{2}\eta(t) \qquad \text{if} \quad 0 \le t \le \varepsilon \tag{20.4}$$

which obviously implies $\eta(t) = 0$ if $0 \le t \le \varepsilon$. For estimating $\eta(t)$, we write

$$\begin{aligned} w &= B(S(t)u_0, w) + B(v - S(t)u_0, w) + \\ &\quad + B(w, u - S(t)u_0) + B(w, S(t)u_0) \\ &= A + B + C + D. \end{aligned} \tag{20.5}$$

For estimating $\|D(t)\|_{3,\infty}$ we return to the definition of the bilinear operator and bound $\left\|S(t-s)\,\partial_j[w_j(s)\,S(s)u_0]\right\|_{3,\infty}$ by $C(t-s)^{-1/2}\,\|w_j\|_{3,\infty} \cdot \|s^{1/2}S(s)u_0\|_\infty\, s^{-1/2}$. This yields the trivial estimate

(20.6)
$$\|B(w, S(t)u_0)\|_{(3,\infty)} \le C\,\pi\,\eta(t) \sup_{0 \le s \le t} \|s^{1/2}\,S(s)u_0\|_\infty .$$

A similar estimate holds for $\|A(t)\|_{(3,\infty)}$. Concerning $\|B(t)\|_{(3,\infty)}$ and $\|C(t)\|_{(3,\infty)}$, we apply lemma 23 and observe that

$$\|u - S(t)u_0\|_{(3,\infty)} \le \|u - S(t)u_0\|_3 = \alpha(t)$$

where $\alpha(t)$ tends to 0 with t.

All together we obtain

$$\|w(\cdot,t)\|_{(3,\infty)} \leq \gamma(t) \sup_{0\leq s\leq t} \|w(\cdot,s)\|_{(3,\infty)} \tag{20.7}$$

where $\gamma(t)$ tends to 0 with t. This immediately implies (20.1) and theorem 20.1 follows.

21 Kato's program : the Morrey-Campanato spaces

We now turn to another application of T. Kato's program : the Morrey-Campanato space $E = M^2(\mathbb{R}^3)$ which is defined by

$$\sup_{x_0\in\mathbb{R}^3} \sup_{R>0} R^{-1} \int_{|x-x_0|\leq R} |f(x)|^2\, dx < \infty\,. \tag{21.1}$$

Keeping the same notations and definitions as above, we observe that $E = F^*$ where F is a separable Banach space. Then $Y = C([0,\infty), E)$ is given the same meaning as in section 12 and $Z \subset Y$ is the subspace of Y consisting of all functions $u(\cdot,t)$ for which

$$\sup_{t\geq 0} \{t^{1/2}\, \|u(\cdot,t)\|_\infty + \|u(\cdot,t)\|_E\} = \|u\|_Z \tag{21.2}$$

is finite.

We already know from theorem that the expected estimate

$$\|B(f,g)\|_Y \leq C\, \|f\|_Y\, \|g\|_Y \tag{21.3}$$

cannot hold. Indeed E is not fully adapted to Navier-Stokes equations.

However, as it was the case when $E = L^3(\mathbb{R}^3)$, it is not difficult to obtain

$$\|B(f,g)\|_Z \leq C\, \|f\|_Z\, \|g\|_Z\,. \tag{21.4}$$

Once more the easy part in proving (21.4) concerns the E component of the Z-norm. Indeed we have

$$\|fg\|_E \leq \|f\|_\infty\, \|g\|_E \leq s^{-1/2}\, \|f\|_Z\, \|g\|_Z \tag{21.5}$$

if $f = f(\cdot, s)$, $g = g(\cdot, s)$. Then (19.224) implies

(21.6)
$$\|B(u,v)\|_E \le \|\omega\|_1 \left(\int_0^t (t-s)^{-1/2}\, s^{-1/2}\, ds \right) \|u\|_Z\, \|v\|_Z$$

where the notations are the sames as in section 19.

We need to treat the L^∞-component of $w = B(u,v)$. Let us denote by $W(s,t,x)$ the integrand $S(t-s)\,\partial_j(u_j(s)v(s))$. We then have

$$\|W(s,t,x)\|_\infty \le C(t-s)^{-1/2}\, s^{-1}\, \|u\|_Z\, \|v\|_Z$$

since $\|u(s)\|_\infty \le s^{-1/2}\, \|u\|_Z$ and the same for v. On the other hand $\|uv\|_E \le \|u\|_\infty\, \|v\|_E$. Finally lemma 9 is again used and yields

$$\|W(s,t,x)\|_\infty \le C(t-s)^{-1}\, s^{-1/2}\, \|u\|_Z\, \|v\|_Z\,.$$

We have obtained two extimates on $\|W(s,t,x)\|_\infty$ and their geometrical mean yields

(21.7)
$$\|W(s,t,x)\|_\infty \le C(t-s)^{-3/4}\, s^{-3/4}\, \|u\|_Z\, \|v\|_Z\,.$$

This can be integrated over $[0,t]$ and we therefore obtain $\|w(\cdot,t)\|_\infty \le Ct^{-1/2}\, \|u\|_Z\, \|v\|_Z$.

The continuity of $B(u,v)$ with respect to the time variable is treated as in section 18 and we obtain

Theorem 21.1. *If $E = M^2(\mathbb{R}^3)$, there exists a positive number η such that for every $u_0 \in E$ with $\|u_0\|_E < \eta$, there exists a solution*

(21.8)
$$u(x,t) \in C([0,\infty); E)$$

of Navier-Stokes equations which satisfies

(21.9)
$$u(x,0) = u_0(x)$$

(21.10)
$$\sup_{t\ge 0} \left(\|u(\cdot,t)\|_E + t^{1/2}\, \|u(\cdot,t)\|_\infty \right) \le C\, \|u_0\|_E\,.$$

Moreover there exists a second positive constant $\beta > 2\eta$ such that this solution is uniquely defined by

$$\sup_{t\ge 0} \left(\|u(\cdot,t)\|_E + t^{1/2}\, \|u(\cdot,t)\|_\infty \right) < \beta\,.$$

This section will be concluded with a remark concerning the preceding examples. For defining mild solutions, we needed to give a specific meaning to the pointwise product $u_j(s)\,u(s)$ between two components of the velocity field.

Either we were assuming that the Banach space E was fully adapted to Navier-Stokes equations. Then we could write $fg = \Lambda h$ for any two functions f, g in E and we knew that h belongs to E. This was the basic property on which the continuity of the bilinear operator B was grounded.

Or we were assuming that E was defined by size conditions on f. This implied $\|fg\|_E \le \|f\|_E\,\|g\|_\infty$ and this observation was used when $E = L^3(\mathbb{R}^3)$ or $E = M^2(\mathbb{R}^3)$.

We now consider a third option and this leads us to the next section.

22 Kato's algorithm revisited

In a joint work with M.A. Muschietti, we planned to apply T. Kato's algorithm to Banach spaces which are defined by regularity conditions instead of size conditions. In such contexts it is no longer possible to multiply a function in E with a function in L^∞. In other words, L^∞ should be replaced by a Banach algebra A of pointwise multipliers of functions of E.

This Banach algebra is given the operator norm of the corresponding multiplier. In other words

$$\|m\|_A = \sup\left\{\|m(x)\,f(x)\|_E\,;\ \|f\|_E \le 1\right\}. \tag{22.1}$$

The relation between A and E is specified by the following assumption.

Definition 20. *Let E be a functional Banach space. Let us assume that E is adapted to the Navier-Stokes equations. Let us assume that for $f \in E$ and $g \in E$ we have*

$$\|fg\|_E \le C\big(\|f\|_\infty + \|\Lambda f\|_E\big)\,\|g\|_E\,. \tag{22.2}$$

We finally assume that the Riesz transformations are bounded on E. We then say that E is adapted to Kato's algorithm.

Property (22.2) can be rephrased into

$$\|f\|_A \le C\big(\|f\|_\infty + \|\Lambda f\|_E\big) \tag{22.3}$$

where $\|f\|_A$ is the pointwise multiplier norm. We then have

Theorem 22.1. *Let us assume that the Banach space E is adapted to Kato's algorithm.*

Then there exists a positive number η such that for every $u_0 \in E$ with

$$\|u_0\|_E < \eta \quad , \quad \operatorname{div} u_0 = 0 \tag{22.4}$$

there exists a mild solution $u \in C([0,\infty); E)$ to the Navier-Stokes equations such that $u(x,0) = u_0(x)$.

This solution satisfies

$$\sup_{t \geq 0} \left(\|u(\cdot,t)\|_E + t^{1/2} \|u(\cdot,t)\|_A \right) \leq C \|u_0\|_E \tag{22.5}$$

where $C = C(E)$ is a constant.

Finally there exists a second positive number $\beta > 2\eta$ such that this solution is uniquely defined by

$$\sup_{t \geq 0} \left(\|u(\cdot,t)\|_E + t^{1/2} \|u(\cdot,t)\|_A \right) < \beta .$$

The proof of theorem 22.1 is similar in spirit to Kato's approach.

Let us define the Z-norm of $u(\cdot,t)$ as

$$\|u\|_Z = \sup_{t \geq 0} \left\{ \|u(\cdot,t)\|_E + t^{1/2} \|u(\cdot,t)\|_A \right\} \tag{22.6}$$

and show the following properties

$$\|B(u,v)\|_Z \leq C \|u\|_Z \|v\|_Z \tag{22.7}$$

$$\|S(t)u_0\|_Z \leq C \|u_0\|_E . \tag{22.8}$$

Then lemma 20 will be applied and theorem 22.1 will follow.

The proof of (22.261) mimics the one we gave when $E = L^3(\mathbb{R}^3)$. Indeed

$$\begin{aligned} \|B(u,v)\|_E &\leq C \int_0^t \|S(t-s)\, \partial_j(u_j v)\|_E \, ds \\ &\leq C' \left(\int_0^t (t-s)^{-1/2} s^{-1/2} \, ds \right) \|u\|_Z \|v\|_Z = C'\pi \|u\|_Z \|v\|_Z . \end{aligned}$$

On the other hand $\|S(t-s)\, \partial_j(u_j v)\|_A \leq C(t-s)^{-1/2} s^{-1} \|u\|_Z \|v\|_Z$ since A is a Banach algebra.

For obtaining an other estimate, we apply (22.257). We have $\|S(t-s)$ $\partial_j(u_j u)\|_\infty \leq C(t-s)^{-1} \|u_j v\|_E \leq C(t-s)^{-1} s^{-1/2} \|u\|_Z \|v\|_Z$. We obtained the first estimate by lemma 9 and the second by $\|u_j v\|_E \le \|u\|_A \, \|v\|_E$. On the other hand $\|\Lambda S(t-s)\, \partial_j(u_j u)\|_E \le C(t-s)^{-1} \, \|u_j v\|_E \le C(t-s)^{-1}$ $s^{-1/2} \, \|u\|_Z \, \|v\|_Z$.

Then (22.3) yields $\|S(t-s)\, \partial_j(u_j v)\|_A \le C(t-s)^{-1}\, s^{-1/2} \, \|u\|_Z \, \|v\|_Z$. This, together with the first estimate, implies

$$\|S(t-s)\, \partial_j(u_j v)\|_A \le C(t-s)^{-3/4}\, s^{-3/4} \, \|u\|_Z \, \|v\|_Z \tag{22.9}$$

which yields $\|B(u,v)\|_A \le C t^{-1/2} \, \|u\|_Z \, \|v\|_Z$.

Everything runs smoothly and it suffices to study the linear evolution $S(t)u_0$.

We apply (22.3) to bound $\|S(t)u_0\|_A$. We need to bound $\|S(t)u_0\|_\infty$ together with $\|\Lambda S(t)u_0\|_E$. Everything works as above and we obtain the required estimate.

For later use we state the following remark.

Lemma 26. *If $m(x)$ belongs to the multiplier algebra A, so does $m(\lambda x)$ for $\lambda > 0$ and we have*

$$\|m(\lambda x)\|_A = \|m(x)\|_A \,. \tag{22.10}$$

Indeed $\|m(\lambda x)\|_A = \sup\left\{\|m(\lambda x)\, f(x)\|_E \,;\, \|f\|_E \le 1\right\}$. We write $f(x) = \lambda g(\lambda x)$ and $\|f\|_E \le 1$ is equivalent to $\|g\|_E \le 1$. Then

$$m(\lambda x)\, f(x) = \lambda m(\lambda x)\, g(\lambda x)$$

implies $\|m(\lambda x) f(x)\|_E = \|m(x)\, g(x)\|_E$. This immediately implies (22.10).

23 Improved regularity of solutions of Navier-Stokes equations

We now answer the problem of obtaining an improved regularity for mild solutions to Navier-Stokes equations. This improved regularity only concerns the behavior of $u(x,t)$ for $t > 0$. We return to the definition of the Banach space Z which is used to prove the convergence of the iterative scheme. We now modify the definition of the Z-norm and write

(23.1)
$$\|u(\cdot,t)\|_Z = \sup_{t\ge 0} \left\{\|u(\cdot,t)\|_E + t^{1/2} \, \|\Lambda u(\cdot,t)\|_E + t^{1/2} \, \|u(\cdot,t)\|_A\right\}.$$

We already proved that the bilinear operator B is continuous for this modified Z-norm. Indeed $t^{1/2}\,\|\Lambda u(\cdot,t)\|_E$ was used for controlling $t^{1/2}\,\|u(\cdot,t)\|_A$. Then the proof of theorem 22.1 tells us the following : the solution which is built through the iterative scheme satisfies $t^{1/2}\,\|\Lambda u(\cdot,t)\|_E \le C < \infty$.

We now check that this improved regularity in the x variable implies an improved regularity in the t variable. Indeed we have

Lemma 27. *For $0 \le \alpha \le 1$ and $t \ge 0$ we have*

$$\|(S(t)-I)f\|_E \le C\,t^{\alpha}\,\|\Lambda^{2\alpha} f\|_E \tag{23.2}$$

for every f in E.

On the Fourier transform side, it amounts to checking that $\frac{e^{-t|\xi|^2}-1}{t^{\alpha}\,|\xi|^{2\alpha}}$ is the Fourier transform of $\frac{1}{t^{3/2}}\,g_\alpha\big(\frac{x}{\sqrt{t}}\big)$ with $g_\alpha \in L^1(\mathbb{R}^3)$. This is trivial when $0 < \alpha \le 1$ and lemma is trivial if $\alpha = 0$.

Let us study the regularity in t of the bilinear operator. If $t' \ge t$, we need to estimate the difference

$$\begin{aligned} &\int_0^{t'} S(t'-s)\,\partial_j(u_j u)\,ds - \int_0^{t} S(t-s)\,\partial_j(u_j u)\,ds \\ &\quad = \int_t^{t'} S(t'-s)\,\partial_j(u_j u)\,ds + \int_0^{t} \{S(t'-s) - S(t-s)\}\,\partial_j(u_j u)\,ds \\ &\quad = U + V\,. \end{aligned}$$

We obviously have $\|U\|_E \le C\int_t^{t'}(t'-s)^{-1/2}\,\|u_j u\|_E\,ds \le C'\big(\int_t^{t'}(t'-s)^{-1/2}\cdot s^{-1/2}\,ds\big)\,|u|_Z^2$. Finally we obtain

$$\|U\|_E \le C'(t'-t)^{1/2}\,t^{-1/2}\,\|u\|_Z^2\,.$$

For controlling $\|V\|_E$, we write

$$S(t'-s) - S(t-s) = [S(t'-t) - I]\,S(t-s)\,. \tag{23.3}$$

We want to apply lemma 27 with $\alpha = 1/2$. We are led to estimating

$$\left\|\int_0^t \Lambda S(t-s)\,\partial_j(u_j u)\,ds\right\|_E\,. \tag{23.4}$$

A key observation is the following estimate

$$\|\partial_j(u_j u)\|_E \le \frac{C}{s}\,\|u\|_Z^2\,. \tag{23.5}$$

Indeed Leibniz' rule is applied and $\|u_j\partial_j u_k\|_E$ is estimated by $\|u_j\|_A \, \|\partial_j u_k\|_E$. We then apply our new definition of the Z-norm and we use the identity $-i\partial_j = R_j\Lambda$ where R_j is the Riesz transformation. All together we obtain (23.5).

Returning to $\|\Lambda S(t-s)\,\partial_j(u_j u)\|_E$ we either use (23.5) and obtain $C(t-s)^{-1/2}\,s^{-1}\,\|u\|_Z^2$ or group together $\Lambda S(t-s)\partial_j$. This second option yields $C(t-s)^{-1}\,\|u_j u\|_E \le C'(t-s)^{-1}\,s^{-1/2}\|u\|_Z^2$. Altogether these two estimates yield $\|\Lambda S(t-s)\,\partial_j(u_j u)\|_E \le C(t-s)^{-3/4}\,s^{-3/4}\,\|u\|_Z^2$ and

$$\left\| \int_0^t \Lambda S(t-s)\,\partial_j(u_j u)\,ds \right\|_E \le C t^{-1/2}\,\|u\|_Z^2 . \tag{23.6}$$

We have proved the following theorem

Theorem 23.1. *Let us assume that the functional Banach space E is adapted to Kato's algorithm. Then the solution $u(x,t)$ to the Navier-Stokes equations which is constructed in theorem 22.1 satisfies*

$$\|u(x,t') - u(x,t)\|_E \le C(u_0)(t'-t)^{1/2}\,t^{-1/2} \tag{23.7}$$

if $t' \ge t$.

This estimate is only interesting if $t \le t' \le 2t$.

We now combine theorem 23.1 to theorem 20.1. They imply the following

Theorem 23.2. *There exists a positive number η with the following property. For every $u_0(x)$ in $L^3(\mathbb{R}^3)$ with $\|u_0\|_3 < \eta$, there exists a unique mild solution $u(x,t) \in C([0,\infty); L^3(\mathbb{R}^3))$ of Navier-Stokes equations such that $u(x,0) = u_0(x)$. Moreover this solution satisfies the improved regularity property*

$$\|u(x,t') - u(x,t)\|_3 \le C\left|\frac{t'}{t} - 1\right|^{1/2} \tag{23.8}$$

if $0 < t < t' < 2t$.

The constant C which appears in (23.8) only depends on $\|u_0\|_3$.

24 Examples of Banach spaces which are adapted to Kato's algorithm

The first group of examples is trivial. If E is a functional Banach space and if any function $m(x)$ which belongs to $L^\infty(\mathbb{R}^3)$ is a pointwise multiplier of E, then E is adapted to Kato's algorithm.

This remark applies to $L^3(\mathbb{R}^3)$, $L^{3,\infty}(\mathbb{R}^3)$ and more generally to the Lorentz spaces $L^{3,q}(\mathbb{R}^3)$ for $1 \le q \le \infty$. Another example is the Morrey-Campanato space M^2.

The second group consists of the Besov spaces $B_q = \dot{B}_q^{\alpha,\infty}$ where $\alpha = 3/q - 1$. But here we will not demand the condition $1 \le q < 3$. Indeed we have

Theorem 24.1. *If $1 \le q < 6$, the homogeneous Besov space $\dot{B}_q^{\alpha,\infty}$ is adapted to Kato's algorithm.*

We need to show that $m(x) \in L^\infty(\mathbb{R}^3)$ and $\Lambda m \in B_q$ imply that $m(x)$ is a pointwise multiplier for B_q. The paraproduct algorithm is used and we obtain

$$m(x)\, f(x) \;=\; A(x) + B(x) + C(x)$$

where

$$A(x) \;=\; \sum_{-\infty}^{\infty} S_{j-2}(m)\, \Delta_j(f)$$

$$B(x) \;=\; \sum_{-\infty}^{\infty} S_{j-2}(f)\, \Delta_j(m)$$

and

$$C(x) \;=\; \sum_{|j'-j|\le 2}\sum \Delta_j(m)\, \Delta_{j'}(f)\,.$$

Concerning $A(x)$ everything is trivial since $\|S_j(m)\|_\infty \le C\|m\|_\infty$. The series is a Littlewood-Paley expansion and A belongs to B_q.

The second series is similar. We have $\|S_j(f)\|_\infty \le C2^j\|f\|_{B_q}$ while $\|\Delta_j(m)\|_q \le C\,2^{-3j/q}$ follows from $\Lambda m \in B_q$.

As always the devil is hidden in the third series. We first begin with the trivial case when $1 \le q < 3$. Then $\|\Delta_j(m)\, \Delta_{j'}(f)\|_q \le C\|m\|_\infty\, \|\Delta_{j'}(f)\|_q \le C\|m\|_\infty\, 2^{j(1-3/q)}$ since $|j' - j| \le 2$. Finally lemma 18 applies and yields $C \in B_q$.

When $3 \le q < 6$, this approach fails and $\Lambda m \in B_q$ is needed. Indeed we have $\|\Delta_j(m)\|_q \le C\,2^{-3j/q}$ which implies $\|\Delta_j(m)\, \Delta_{j'}(f)\|_{q/2} \le C\,2^{j(1-6/q)}$.

Since $q < 6$, lemma 18 can be used and yields $C(x) \in \dot{B}_{q/2}^{6/q-1,\infty}$. Now it is trivial to check that $\dot{B}_{q/2}^{6/q-1,\infty}$ is contained in $\dot{B}_q^{3/q-1,\infty}$ for $1 \le q \le \infty$. The same proof works as well if $\dot{B}_1^{2,\infty}$ is replaced by the minimal space $\dot{B}_1^{2,1}$.

Let us now observe that theorem 23.2 is sharp. Indeed $\dot{B}_6^{-1/2,\infty}$ is not adapted to Kato's algorithm.

Here is a simple counter-example. We define $m(x) = \left(\sum_0^\infty 2^{-j}\, e^{i4^j x_1}\right)\varphi(x)$ where the Fourier transform of φ is supported by $|\xi| \le 1/2$. We then have $m(x) \in L^\infty(\mathbb{R}^3)$ and $\Lambda m \in \dot{B}_q^{-1/2,\infty}$ for $1 \le q \le \infty$. On the other hand $f(x) = \left(\sum_0^\infty 2^j\, e^{-i4^j x_1}\right)\varphi(x)$ belongs to $\dot{B}_6^{-1/2,\infty}$ (indeed to $\dot{B}_q^{-1/2,\infty}$ for $1 \le q \le \infty$) while $m(x)\, f(x)$ does not belong to $\mathcal{S}'(\mathbb{R}^3)$. In fact the paraproduct algorithm applies to $m(x)\, f(x)$. The series which were denoted by $A(x)$ and $B(x)$ trivially belong to $\dot{B}_6^{-1/2,\infty}$ while $C(x) = \sum_0^\infty \varphi^2(x)$ does not converge in the distributional sense.

Does it mean that Navier-Stokes equations cannot be solved through Kato's algorithm when $E = \dot{B}_6^{-1/2,\infty}$? M. Cannone addressed this issue [12] and proposed the following solution.

The Banach space Z is now defined by the following two conditions

$$\sup_{t \ge 0} \|u(\cdot, t)\|_E < \infty \tag{24.1}$$

together with

$$\sup_{t \ge 0} t^{1/4}\, \|u(\cdot, t)\|_6 < \infty . \tag{24.2}$$

It is then an easy exercise to check the following two estimates on the bilinear term. We do have

$$\|B(u, v)\|_3 \le C\, \|u\|_Z\, \|v\|_Z \tag{24.3}$$

and

$$t^{1/4}\, \|B(u, v)\|_6 \le C\, \|u\|_Z\, \|v\|_Z . \tag{24.4}$$

We then observe that $L^3(\mathbb{R}^3)$ is continuously embedded in $\dot{B}_6^{-1/2,\infty}$ and lemma 20 can be applied.

Returning to theorem 22.1, an interesting application is given by the homogeneous Sobolev space $\dot{H}^{1/2}$ which coincides with the homogeneous Besov space $\dot{B}_2^{1/2,2}$.

If $m(x)$ belongs to $L^\infty(\mathbb{R}^3)$, the series of the series $A(x) = \sum_{-\infty}^\infty S_{j-2}(m) \cdot \Delta_j(f)$ belongs to $\dot{H}^s(\mathbb{R}^3)$ whenever f does. The value of s is irrelevant.

On the other hand if Λm belongs to $\dot{H}^{1/2}$, then m belongs to $\dot{H}^{3/2}(\mathbb{R}^3)$ and

$$B(x) = \int_{-\infty}^{\infty} S_{j-2}(f)\,\Delta_j(m) \qquad \text{belongs to} \quad H^{1/2}.$$

Indeed we have $\|S_j(f)\|_\infty \le C\,2^j$ and $\|\Delta_j(m)\|_2 \le \varepsilon_j\,2^{-3/2j}$ with $\sum_{-\infty}^{\infty} \varepsilon_j^2 < \infty$.

The third series is $C(x) = \sum\sum_{|j'-j|\le 2} \Delta_j(f)\,\Delta_{j'}(m)$. We have $\|\Delta_j(f)\cdot \Delta_{j'}(m)\|_1 \le \eta_j\,2^{-2j}$ where $\eta_j \in \ell^1(\mathbf{Z})$. Lemma 18 can therefore be applied and yields $C(x) \in \dot{B}_1^{2,1}$ which is the minimal Banach space described in section 8.

Since $\dot{H}^{1/2}$ is embedded into $L^3(\mathbb{R}^3)$ the proof we gave for the uniqueness of mild solutions applies as well. We therefore obtain the following

Theorem 24.2. *There exists a positive number η with the following property:*

> *for each $u_0(x)$ in the homogeneous Sobolev space $\dot{H}^{1/2}(\mathbb{R}^3)$ fulfilling $\|u_0\|_{\dot{H}^{1/2}} < \eta$ there exists a unique global solution $u(x,t) \in C([0,\infty); \dot{H}^{1/2}(\mathbb{R}^3))$ of Navier-Stokes equations such that $u(x,0) = u_0(x)$.*

We obtain a similar statement if $\dot{H}^{1/2}(\mathbb{R}^3)$ is replaced by the minimal Banach space $\dot{B}_1^{2,1}$.

This minimal space is interesting since it answers a natural question concerning regularity. The largest amount of regularity which can be discussed inside the framework of adapted Banach spaces is precisely defined by the $\dot{B}_1^{2,1}$ norm. We cannot go further and deal with three derivatives. The corresponding Banach space would not be adapted to Navier-Stokes equations.

On the other hand it is fortunate that our modification of Kato's algorithm applies to this minimal space $\dot{B}_1^{2,1}$.

It is also clear that a wavelet analysis yields the best understanding of this minimal space $\dot{B}_1^{2,1}$. Indeed wavelets provide us with the canonical isomorphism between $\dot{B}_1^{2,1}$ and ℓ^1.

From these remarks one can argue that wavelets still have some links with Navier-Stokes equations.

25 Self-similar solutions of Navier-Stokes equations

The existence of self-similar solutions of the Navier-Stokes equations is a remarkable application of the results which were presented in the preceding section.

A Banach space E is assumed to be adapted to Kato's algorithm and is also assumed to contain non-trivial homogeneous functions of degree -1. An example is given by $E = L^{3,\infty}$ or $E = \dot{B}_1^{2,\infty}$ or by the Morrey-Campanato space M^2.

A self-similar solution of Navier-Stokes equations is a solution $u \in C([0,\infty); E)$ which is invariant under the canonical rescaling. In other words

$$(25.1) \qquad u(x,t) = \lambda u(\lambda x, \lambda^2 t)\,, \qquad 0 < \lambda < \infty\,.$$

It implies

$$(25.2) \qquad u(x,t) = \frac{1}{\sqrt{t}}\,U\Big(\frac{x}{\sqrt{t}}\Big) \quad , \quad U(x) = u(x,1)\,.$$

We then necessarily have

$$(25.3) \qquad \lambda u_0(\lambda x) = \lim_{t \downarrow 0} \lambda u(\lambda x, \lambda^2 t) = u_0(x)\,.$$

Let us conversely assume $u_0 \in E$, $\lambda u_0(\lambda x) = u_0(x)$, $0 < \lambda < \infty$ and $\|u_0\|_E < \eta$ where η is a positive constant whose value will be given in the proof.

We want to show that the corresponding solution of Navier-Stokes equations which is given by theorem 22.1 is indeed a self-similar solution.

Let us denote by $u_\lambda(x,t)$ the function $\lambda u(\lambda x, \lambda^2 t)$. We already know that $u_\lambda(x,t)$ is a solution of the Navier-Stokes equations. Next we observe that $u_\lambda(x,0) = u(x,0) = u_0(x)$.

Finally we need to show that u_λ satisfies (22.5) when u does. Indeed $\sup_{t\geq 0} \|u_\lambda(\cdot,t)\|_E = \sup_{t\geq 0} \|u(\cdot,t)\|_E$ while, for every $m \in A$

$$(25.4) \qquad \|m(\lambda x)\|_A = \|m(x)\|_A\,.$$

This immediately yields

$$\sup_{t\geq 0} t^{1/2}\,\|u_\lambda(\cdot,t)\|_A = \sup_{t\geq 0} \lambda t^{1/2}\,\|u(\lambda x, \lambda^2 t)\|_A$$
$$= \sup_{t\geq 0} t^{1/2}\,\|u(\lambda x, t)\|_A = \sup_{t\geq 0} t^{1/2}\,\|u(x,t)\|_A\,.$$

Our claim is now proved and the uniqueness of the solution constructed in theorem 22.1 yields $u_\lambda = u$. We just proved the following result.

Theorem 25.1. *Let us assume that a functional Banach space E is adapted to Kato's algorithm as indicated in definition 20.*

If an initial condition $u_0(x)$ belongs to E with $\|u_0\|_E < \eta$ and satisfies the homogeneity condition

$$\lambda u_0(\lambda x) = u_0(x)\,, \qquad 0 < \lambda < \infty\,, \tag{25.5}$$

then the corresponding solution $u(x,t)$ of Navier-Stokes equations is a self-similar solution.

As it was specified, $\eta > 0$ is the constant that appears in theorem 22.1 and the solution which is here refered to is the unique solution of Navier-Stokes equation described in theorem 22.1.

This general theorem covers and extends some of the *ad-hoc* results which were obtained in [12] or [13].

Let us be more specific about a special case. If E is the Lorentz space $L^{3,\infty}(\mathbb{R}^3)$, then the sophistication of Kato's algorithm which is given in section 22 is not needed. Theorem 18.2 implies the existence of self-similar solutions f in $L^{3,\infty}(\mathbb{R}^3)$.

Moreover the subspace of $L^{3,\infty}(\mathbb{R}^3)$ which is defined by $f(\lambda x) = \lambda^{-1}\cdot f(\lambda x)$, $0 < \lambda < \infty$, is extremely simple. Indeed $f(x)$ belongs to this subspace if and only if the restriction of $f(x)$ to the unit sphere S^2 belongs to $L^3(S^2)$.

This means that the search for self-similar solutions belonging to $C([0,\infty); L^{3,\infty}(\mathbb{R}^3))$ is a simple and natural problem which can be treated in a self-consistent approach.

There is however an interesting construction of self-similar solutions which is not covered by our general approach.

Indeed in [13] we fix a large integer m and consider the homogeneous space $\dot{E}_m$ defined by $|\partial^\alpha f(x)| \le C|x|^{-1-|\alpha|}$, $|\alpha| \le m$. This Banach space is not translation invariant and cannot be incorporated inside our general approach. However if $u_0(x)$ satisfies $\lambda u_0(\lambda x) = u_0(x)$, $0 < \lambda < \infty$, and if the norm of $u_0(x)$ in $\dot{E}_m$ is small, then the corresponding solution $u(x,t)$ of Navier-Stokes equations belongs to $C([0,\infty); \dot{E}_m)$.

This theorem which is proved in [13] cannot be covered by our general theory. Indeed inside this general theory, the regularity cannot exceed 2 since $\dot{B}_1^{2,1}(\mathbb{R}^3)$ is the minimal adapted space.

26 Uniqueness of mild solutions to Navier-Stokes equations

We return to the remarkable theorem by G. Furioli, P.G. Lemarié-Rieusset and E. Terraneo.

In section 20 was given a simple proof of this discovery (theorem 20.1).

Here we would like to present the original proof which is based on paraproduct estimates or spectral methods. In contrast our new proof depends on real variable estimates and we cannot state as we did before that Littlewood-Paley analysis is actually needed for proving uniqueness.

In this section, all the results of section 20 are forgotten and the proof of uniqueness starts from scratch.

Therefore section 20 does not exist and this section is viewed as a continuation of section 19. We begin with recalling what was proved in section 19.

We return to the case where the adapted Banach space E is $L^3(\mathbb{R}^3)$ and study mild solutions $u(x,t) \in C([0,\infty); E)$ to Navier-Stokes equations. Existence of such solutions was proved under the condition $\|u_0\|_3 < \eta$ and uniqueness under the condition

$$\sup_{t\geq 0}\left\{\|u(\cdot,t)\|_3 + t^{1/2}\,\|u(\cdot,t)\|_\infty\right\} < \beta \tag{26.1}$$

where $\beta > \eta$ is a small constant.

We still do not know if the assumption $\|u_0\|_3 < \eta$ is actually needed. But P.G. Lemarié-Rieusset proved uniqueness without assuming (26.1). We present here this remarkable theorem [35].

Theorem 26.1. *For $T > 0$, let $u(x,t)$ and $v(x,t)$ belong to $C([0,T]; L^3)$, $L^3 = L^3(\mathbb{R}^3)$.*

If u and v are two mild solutions of Navier-Stokes equations such that $u(\cdot,0) = v(\cdot,0) = u_0(x)$, then $u(\cdot,t) = v(\cdot,t)$ for $0 \leq t \leq T$.

The proof relies on some new estimates on the bilinear operator $B(f,g)$.

Let E be the homogeneous Besov space $\dot{B}_2^{1/2,\infty}(\mathbb{R}^3)$. Then E is not embedded in $L^3(\mathbb{R}^3)$ and conversely $L^3(\mathbb{R}^3)$ is not embedded in E.

As it was stated before, the bilinear operator

$$B(f,g) = \int_0^t S(t-s)\,\Lambda\big[f(\cdot,s)\,g(\cdot,s)\big]\,ds$$

is not bounded from $Y \times Y$ into Y when $Y = C([0,\infty); L^3(\mathbb{R}^3))$.

However we have

Proposition 1. *There exists a constant C such that for any $f(x,t)$, any $g(x,t)$ and $t \geq 0$*

$$\|B(f,g)\|_E(t) \leq C \sup_{0\leq s\leq t} \|f(\cdot,s)\|_3 \sup_{0\leq s\leq t} \|g(\cdot,s)\|_3 \,. \tag{26.2}$$

This result will be completed with the following ones

Proposition 2. *With the same notations, we have*

$$\|B(f,g)\|_E(t) \leq C \sup_{0\leq s\leq t} \|f(\cdot,s)\|_3 \sup_{0\leq s\leq t} \|g(\cdot,s)\|_E \,. \tag{26.3}$$

Proposition 3. *If $3 \leq q < 6$, there exists a constant C_q such that*

$$\|B(f,g)\|_E(t) \leq C_q \sup_{0\leq s\leq t} \left\{ s^{1/2(1-3/q)} \|f(\cdot,s)\|_q \right\} \cdot \sup_{0\leq s\leq t} \|g(\cdot,s)\|_E \,. \tag{26.4}$$

Let us begin with proposition 1.

We first sketch the proof and then give a complete proof.

The main observation about the bilinear operator $B(f,g)$ concerns the nature of the divergence. As it was already checked many times, the integrand blows up as s reaches t. To get a better understanding of this divergence, one splits this integral into a series

$$\sum B_j(f,g) \quad \text{where} \quad B_j(f,g) = \int_{\{4^{-j-1}\leq t-s\leq 4^{-j}\}} S(t-s)\,\Lambda(fg)\,ds\,.$$

We now mimic $\Lambda S(t-s)$ by $2^j\Delta_j$, we pretend that $f(\cdot,s)$ and $g(\cdot,s)$ are constant functions as $4^{-j-1} \leq t-s \leq 4^{-j}$ and finally replace ds by the step size 4^{-j}.

With such claims, the bilinear operator is modelled by $\sum_{-\infty}^{\infty} 2^{-j}\,\Delta_j(f_j g_j)$ where $\sup_j \|f_j\|_3 < \infty$, $\sup_j \|g_j\|_3 < \infty$ and Δ_j is defined in the section devoted to the Littlewood-Paley analysis. We then have

Lemma 28. *If $\|f_j\|_3 \le C_0$ and $\|g_j\|_3 \le C_0$, then $\sum_{-\infty}^{\infty} 2^{-j}\,\Delta_j(f_j g_j)$ belongs to the homogeneous Besov space $\dot{B}_{3/2}^{1,\infty}$.*

The proof is trivial by lemma 18, section 15. Finally $\dot{B}_{3/2}^{1,\infty}$ is contained in $\dot{B}_2^{1/2,\infty}$.

The actual proof is following the same strategy. The integral which is defining $B(f,g)$ is split into a series of terms where $s \in [t-4^{-j}\,,\, t-4^{-j-1}]$ which yields $w = B(f,g) = \sum_{-\infty}^{\infty} w_j$. It is now easy to control the $L^{3/2}$ norm of w_j as well as the $L^{3/2}$-norm of the second derivatives $\partial^\alpha w_j$, $|\alpha| = 2$. Then one applies lemma 18 or one of its variants.

We would like to give an other proof which was found by M. Cannone. We start with the following observation.

Lemma 29. *On the homogeneous Besov space $\dot{B}_3^{-1,1}$, the norm $\sum_{-\infty}^{\infty} 2^{-j}\cdot \|\Delta_j(f)\|_3$ and the continuous counterpart $\int_0^\infty \|\Lambda S(t) f\|_3\,dt$ are equivalent ones.*

Indeed one writes $f = \Lambda g$ where $g \in \dot{B}_3^{0,1}$ and $\Lambda = (-\Delta)^{1/2}$. Then $\Lambda\, S(t) f = -t\Delta \exp(t\Delta)\,\frac{1}{t}$ and $\int_0^\infty \|\Lambda S(t) f\|_3\,dt = \int_0^\infty \|Q(t) g\|_3 \frac{dt}{t}$ where $Q(t) = -t\Delta \exp(t\Delta)$. We therefore obtain a classical definition of the norm of g in the homogeneous Besov space $\dot{B}_3^{0,1}$.

For computing the norm of $B(f,g)$ in $\dot{B}_{3/2}^{1,\infty}$, it suffices to estimate

$$\sup\left\{ < B(f,s), h > ;\ \|h\|_{\dot{B}_3^{-1,1}} \le 1 \right\}.$$

One writes $h(x,t) = \Lambda\, S(t) h$ and obtains

$$< B(f,g), h > = \int_0^t \int_{\mathbb{R}^3} h(x, t-s)\, f(x,s)\, g(x,s)\, dx\, ds\,.$$

One is then using lemma 29 for obtaining

$$\begin{aligned} |< B(f,g), h >| &\le \sup_{0\le s\le t} \|f(\cdot,s)\|_3 \sup_{0\le s\le t} \|g(\cdot,s)\|_3 \int_0^\infty \|h(\cdot,s)\|_3\, ds \\ &\le \sup_{0\le s\le t} \|f(\cdot,s)\|_3 \sup_{0\le s\le t} \|g(\cdot,s)\|_3 \end{aligned}$$

as announced.

The proof of proposition 2 is slightly more involved. Let us observe that proposition 2 is a special case of proposition 3 when $q = 3$. We now concentrate on proposition 3. We begin with rephrasing theorem 15.2 as a lemma.

Lemma 30. *If f and g are two functions defined on $\mathbb{R}^3$, we then have when* $3/2 < q < 6$

$$\|\Delta_j(fg)\|_2 \le C_q\, 2^{j(3/q-1/2)}\, \|f\|_q\, \|g\|_{\dot{B}_2^{1/2,\infty}}\,.$$

For proving (26.4) we write $\Gamma(t) = \Lambda\, S(t)$ with $\Lambda = \sqrt{-\Delta}$ and we want to estimate

$$\sup_{j\in\mathbf{Z}} \left\{ 2^{j/2} \left\| \Delta_j \int_0^t \Gamma(t-s)\, [f(s)\, g(s)]\, ds \right\|_2 \right\}$$

where $f(s) = f(\cdot, s)$, $g(s) = g(\cdot, s)$.

We begin with a trivial observation. One trivially checks that $\Delta_j = \Delta_j(\Delta_{j-1} + \Delta_j + \Delta_{j+1})$.

We concentrate on Δ_j^2 since the two other terms are similar to this one.

This leads to writing $\Delta_j\Gamma(t) = \Gamma_j(t)\Delta_j$ where $\Gamma_j(t) = \Gamma(t)\Delta_j$. We need a trivial remark.

Lemma 31. *The operator norm of $\Gamma_j(t) : L^2(\mathbb{R}^3) \to L^2(\mathbb{R}^3)$ does not exceed $C\, 2^j(1+4^j t)^{-2}$, $j \in \mathbf{Z}$, $t \ge 0$ where C is a constant.*

This remark is trivial if we move to the Fourier transform side. The operator norm of $\Gamma_j(t)$ is the L^∞ norm of the corresponding multiplier which reads

$$|\xi| \exp\left(-t|\xi|^2\right) \hat{\psi}(2^{-j}\xi)\,. \tag{26.5}$$

This L^∞ norm is better computed after performing a dilation $\xi \to 2^j\xi$ and the computation is trivial.

Returning to (26.4) we assume

$$\sup_{0\le s\le t} s^{1/2(1-3/q)}\, \|f(\cdot, s)\|_q \le 1 \tag{26.6}$$

together with

$$\sup_{0\le s\le t} \|g(\cdot, s)\|_E \le 1\,.$$

Then lemma 31 and proposition 3 yield

$$\left\| \Delta_j \int_0^t \Gamma(t-s)\,[f(s)\,g(s)]\,ds \right\|_2 \leq C_q\, 2^{j(1/2+3/q)} \int_0^t [1+4^j(t-s)]^{-2}\, s^{-1/2(1-3/q)}\, ds\,. \tag{26.7}$$

Since $\alpha = \frac{1}{2}(1-\frac{3}{q}) \in [0,1)$, we have

$$\sup_{T\geq 0} \int_0^T [1+(T-\tau)^2]^{-1}\,\tau^{-\alpha}\,d\tau \;=\; C_\alpha \;<\; \infty\,.$$

This estimate is applied to (26.7) with $T = 4^j t$ and $\tau = 4^j s$. Then (26.4) is proved.

We now return to theorem 26.1.

The initial condition $u_0(x)$ belongs to $L^3(\mathbb{R}^3)$ but we do not assume $\|u_0\|_3 < \eta$ where $\eta > 0$ is the small positive number which played a crucial role in theorem 19.1 or in theorem 22.1.

We denote by $u(\cdot,t)$ and $v(\cdot,t)$ two mild solutions of Navier-Stokes equations. Both u and v belong to $C([0,T];L^3(\mathbb{R}^3))$ and $u(\cdot,0) = v(\cdot,0) = u_0(x)$.

We want to show that $u(\cdot,t) = v(\cdot,t)$ if $0 \leq t \leq T$. For proving this fact it suffices to show that there exists a positive τ such that $u(\cdot,t) = v(\cdot,t)$ if $0 \leq t \leq \tau$.

Indeed let T_0 be the upper bound of the set of $\tau > 0$ for which $u = v$ on $[0,\tau]$. Since u and v belong to $C([0,T];L^3(\mathbb{R}^3))$ we have $u = v$ on $[0,T_0]$.

We then consider $\tilde{u}(x,t) = u(x\,,\,T_0+t)$ and $\tilde{v}(x,t) = v(x\,,\,T_0+t)$. These new functions are still mild solutions of Navier-Stokes equations (13.2) if $T_0 < T$. We have $\tilde{v}(x,0) = \tilde{u}(x,0)$. Therefore there exists a positive τ such that $\tilde{u}(x,t) = \tilde{v}(x,t)$ if $0 \leq t \leq \tau$ which contradicts the definition of T_0. Therefore $T_0 = T$. For proving that such a positive τ exists, we write $v = u + w$ which yields

$$w \;=\; B(u+w\,,\,u+w) - B(u,u) \;=\; B(v,w) + B(w,u)\,. \tag{26.8}$$

First we observe that $w(\cdot,t)$ belongs to $C([0,T];B)$ where $B = \dot{B}_2^{1/2,\infty}$. This follows from proposition 2. Let us warm the reader that $u(\cdot,t)$ does not belong to $C([0,T];B)$. Indeed the linear evolution $S(t)u_0$ cannot have this property since $L^3(\mathbb{R}^3)$ is not contained in B. This is a striking difference with our new proof where $L^{3,\infty}$ is used instead of B.

We then write $\alpha(t) = \|w(\cdot,t)\|_B$ and want to prove $\alpha(t) = 0$ on $[0,\tau]$.

Returning to (26.8), we further split the right-hand side into

$$w(\cdot,t) = B\big(v - S(t)v_0, w\big) + B\big(s(t)v_0, w\big) + \\ + B\big(w, u - S(t)u_0\big) + B\big(w, S(t)u_0\big) \tag{26.9}$$

with $u_0 = v_0$.

We then define

$$\varepsilon(t) = \|u - S(t)u_0\|_3 + \|v - S(t)v_0\|_3 \tag{26.10}$$

and we know that $\lim_{t\downarrow 0} \varepsilon(t) = 0$ since u and v both belong to $C([0,T]; L^3(\mathbb{R}^3))$. We similarly define

$$\eta(t) = t^{1/8}\, \|S(t)u_0\|_4 \tag{26.11}$$

and we know that $\eta(t) \le C\|u_0\|_3$ together with

$$\lim_{t\downarrow 0} \eta(t) = 0\,. \tag{26.12}$$

Writing in a systematic way $\overline{f}(t) = \sup\{f(s)\,;\, 0 \le s \le t\}$ when f is a non negative function, we want to prove

$$\overline{\alpha}(t) \le C\,\overline{\alpha}(t)\,[\overline{\varepsilon}(t) + \overline{\eta}(t)]\,. \tag{26.13}$$

This obviously implies

$$\alpha(t) = 0 \qquad \text{on} \quad [0,\tau] \tag{26.14}$$

since $\alpha(t)$ is bounded and $\lim_{t\downarrow 0}\big(\overline{\varepsilon}(t) + \overline{\eta}(t)\big) = 0$. Returning to (26.9) it suffices to apply proposition 3 to each one of the four terms in (26.9).

27 Appendix : construction of a divergence-free wavelet basis

As it was explicitely stated in section 6, the first construction of a divergence-free orthonormal wavelet basis was achieved by G. Battle and P. Federbush. Then P.G. Lemarié-Rieusset built a new basis with better spectral properties. We closely follow Lemarié-Rieusset's paper [50] in this appendix.

We start with the classical construction of the wavelet basis of the form $2^{j/2}\,\psi(2^j t-k)$, $j \in \mathbf{Z}$, $k \in \mathbf{Z}$, $\psi \in \mathcal{S}(\mathbb{R})$.

For building this basis, one begins with a scaling function $\varphi \in \mathcal{S}(\mathbb{R})$ such that

$$\hat{\varphi}(\xi) = 1 \quad \text{on} \quad \left[-\frac{2\pi}{3}, \frac{2\pi}{3}\right] \tag{27.1}$$

$$\hat{\varphi}(\xi) = 0 \quad \text{if} \quad |\xi| \geq \frac{4\pi}{3} \tag{27.2}$$

$$0 \leq \hat{\varphi}(\xi) \leq 1 \ , \quad \hat{\varphi}(\xi) = \hat{\varphi}(-\xi) \tag{27.3}$$

$$|\hat{\varphi}(\pi+\eta)|^2 + |\hat{\varphi}(\pi-\eta)|^2 = 1 \ , \qquad |\eta| \leq \frac{\pi}{3} \, . \tag{27.4}$$

These properties obviously imply

$$\sum_{-\infty}^{\infty} |\hat{\varphi}(\xi+2k\pi)|^2 = 1 \tag{27.5}$$

and $\varphi(t-k)$, $k \in \mathbf{Z}$, is an orthonormal sequence.

Moreover if V_0 denotes the closed linear span of this sequence $\varphi(t-k)$, $k \in \mathbf{Z}$, and if V_j is defined by

$$f(t) \in V_0 \iff f(2^j t) \in V_j \tag{27.6}$$

we have $V_0 \subset V_1$ as (27.1) and (27.2) show.

Then $\bigcap_{-\infty}^{\infty} V_j = \{0\}$, $\bigcup_{-\infty}^{\infty} V_j$ is dense in $L^2(\mathbb{R})$ and one can define W_j as being the orthogonal complement of V_j inside V_{j+1}. We then have

$$f \in W_0 \iff f(2^j t) \in W_j \tag{27.7}$$

and there exists an orthonormal basis of W_0 of the form $\psi(t-k)$, $k \in \mathbf{Z}$. Here ψ belongs to the Schwartz class. Moreover $\hat{\psi}(\xi) = 0$ if $|\xi| \leq 2\pi/3$ or $|\xi| \geq 8\pi/3$.

These properties immediately imply the following

$$2^{j/2}\,\psi(2^j t-k) \ , \quad j \in \mathbf{Z} \ , \ k \in \mathbf{Z} \tag{27.8}$$

is an orthonormal basis of $L^2(\mathbb{R})$.

The first step in the construction is the following lemma

Lemma 32. *There exists an increasing sequence $\tilde{V}_j$ of closed subspaces of $L^2(\mathbb{R})$ such that*

(27.9) $$\bigcap_{-\infty}^{\infty} \tilde{V}_j = \{0\} \quad , \quad \bigcup_{-\infty}^{\infty} \tilde{V}_j \quad \text{is dense in } L^2(\mathbb{R})$$

(27.10) $$f(x) \in \tilde{V}_j \iff f(2x) \in \tilde{V}_{j+1}$$

(27.11) *there exists $\tilde{\varphi} \in \mathcal{S}(\mathbb{R})$ such that $\tilde{\varphi}(x-k)$, $k \in \mathbf{Z}$, is a Riesz basis of $\tilde{V}_0$*

(27.12) $$\frac{d}{dx} : \tilde{V}_j \to V_j \quad \text{is continuous with a dense range.}$$

This lemma will be completed with the following information.

Lemma 33. *There exists a closed subspace $\tilde{W}_j$ of $\tilde{V}_{j+1}$ such that $\tilde{V}_j + \tilde{W}_j = \tilde{V}_{j+1}$ and*

(27.13) $$\frac{d}{dx} : \tilde{W}_j \to W_j$$

is an isomorphism.

This construction is indeed simpler than expected. We firt define $\tilde{\varphi}(x) = \int_x^{x+1} \varphi(t)\, dt$ and lemma 32 is obviously checked.

Similarly $\tilde{\psi}(x)$ is defined by $\int_{-\infty}^{\infty} \psi(t)\, dt$ and lemma 33 follows.

For analyzing $L^2(\mathbb{R}^3)$ we have three options at our disposal. Option 1 is given by the multiresolution analysis $\tilde{V}_j \otimes V_j \otimes V_j$, $j \in \mathbf{Z}$. Option 2 is given by $V_j \otimes \tilde{V}_j \otimes V_j$, $j \in \mathbf{Z}$. Similarly option 3 is provided by $V_j \otimes V_j \otimes \tilde{V}_j$, $j \in \mathbf{Z}$.

When a vector field $u(x) = (u_1(x), u_2(x), u_3(x))$ will be analyzed, option 1 will be used for u_1, option 2 for u_2 and option 3 for u_3.

The frequency channels corresponding to option 1 are denoted bu $W_{j,\alpha}^{(1)}$. Similarly $W_{j,\alpha}^{(2)}$ corresponds to option 2 and $W_{j,\alpha}^{(3)}$ to option 3.

We need to explain the meaning of this index α. Here $\alpha \in A$ where A is the set $\{0,1\}^3 \setminus \{(0,0,0)\}$. In other words $\alpha = (\alpha_1, \alpha_2, \alpha_3)$ where $\alpha_j \in \{0,1\}$ and where $\alpha = (0,0,0)$ is omitted.

Finally $\alpha_1 = 0$ means that $W_{j,\alpha}^{(1)} = \tilde{V}_j \otimes \cdot \otimes \cdot$ while $\alpha_1 = 1$ means $W_{j,\alpha}^{(1)} = \tilde{W}_j \otimes \cdot \otimes \cdot$ and similarly for α_2 and α_3. For instance, if $\alpha = (0,1,1)$, then $W_{j,\alpha}^{(1)} = \tilde{V}_j \otimes W_j \otimes W_j$.

The same notations are used for $W_{j,\alpha}^{(2)}$ and $W_{j,\alpha}^{(3)}$.

Once these notations are being fixed, we return to a vector field $u(x) = (u_1(x), u_2(x), u_3(x))$ and decompose each component into the corresponding wavelet expansion. The first component is treated with the first option and so on. We then have

$$\begin{aligned} u_1(x) &= \sum_\alpha \sum_j w_{j,\alpha}^{(1)}(x) \qquad && w_{j,\alpha}^{(1)} \in W_{j,\alpha}^{(1)} \\ u_2(x) &= \sum_\alpha \sum_j w_{j,\alpha}^{(2)}(x) \qquad && w_{j,\alpha}^{(2)} \in W_{j,\alpha}^{(2)} \\ u_3(x) &= \sum_\alpha \sum_j w_{j,\alpha}^{(3)}(x) \qquad && w_{j,\alpha}^{(3)} \in W_{j,\alpha}^{(3)} . \end{aligned}$$

We now reach the crucial lemma.

Lemma 34. *If $\partial_1 u_1 + \partial_2 u_2 + \partial_3 u_3 = 0$ in the distributional sense, then*

(27.14)
$$\partial_1 w_{j,\alpha}^{(1)} + \partial_2 w_{j,\alpha}^{(2)} + \partial_3 w_{j,\alpha}^{(3)} = 0 , \qquad j \in \mathbf{Z} , \ \alpha \in A .$$

In other words $\operatorname{div} u = 0$ is fully decoupled into frequency channels.

Let us prove this striking property. We denote by $W_{j,\alpha}$ the standard multiresolution frequency channels where the standard multiresolution is $V_j \otimes V_j \otimes V_j$.

Then $f \in W_{j,\alpha}^{(1)}$ implies $\partial_1 f \in W_{j,\alpha}$ by (27.12) or (27.13). Similarly $f \in W_{j,\alpha}^{(2)}$ implies $\partial_2 f \in W_{j,\alpha}$ and the same for the third option. If $\partial_1 u_1 + \partial_2 u_2 + \partial_3 u_3 = 0$ we therefore obtain

(27.15)
$$\sum_j \sum_\alpha (\partial_1 w_{j,\alpha}^{(1)} + \partial_2 w_{j,\alpha}^{(2)} + \partial_3 w_{j,\alpha}^{(3)}) = 0 .$$

We then observe that $L^2(\mathbb{R}^3) = \bigoplus_{j,\alpha} W_{j,\alpha}$ where this sum is orthogonal. Therefore (27.15) is decoupled into (27.14).

The second remark we need is the following.

Lemma 35. *If $\alpha_1 = 1$, then*

(27.16)
$$\partial_1 = W_{j,\alpha}^{(1)} \to W_{j,\alpha}$$

is an isomorphism.

This obviously follows from (27.13).

We now construct a basis for the closed vector space $H_{(j,\alpha)} \subset W^{(j)}_{j,\alpha} \times W^{(2)}_{j,\alpha} \times W^{(3)}_{j,\alpha}$ defined by

$$\partial_1 w^{(1)}_{j,\alpha} + \partial_2 w^{(2)}_{j,\alpha} + \partial_3 w^{(3)}_{j,\alpha} = 0\,. \tag{27.17}$$

Indeed $\alpha \neq (0,0,0)$. Therefore one of the three indices α_1, α_2 or α_3 is 1. We only treat the case $\alpha_1 = 1$ since the other two are fully similar ones.

In the first case (27.17) can be written

$$w^{(1)}_{j,\alpha} = T_1\left(\partial_2 w^{(2)}_{j,\alpha} + \partial_3 w^{(3)}_{j,\alpha}\right). \tag{27.18}$$

The basis we are looking for is given by the following obvious lemma.

Lemma 36. *Let H_1 and H_2 be two Hilbert spaces and $T : H_1 \to H_2$ be a continuous linear mapping.*

Let $V \subset H_1 \times H_2$ be the graph of this mapping T.

Then for each Riesz basis e_j, $j \in J$, of H_1, the collection $\left(e_j, T(e_j)\right)_{j\in J}$ is a Riesz basis of V.

Coming back to (27.18) it suffices to treat the case $j = 0$ since everything is dilation invariant. If $j = 0$, everything is invariant under the $\mathbf{Z}^3$ group action. We are therefore led to constructing the "mother divergence-free wavelets". If, for instance, $\alpha = (1,0,0)$, we then either obtain

$$\left(-\tilde{\psi}(x_1)\left[\varphi(x_2+1) - \varphi(x_2)\right]\varphi(x_3)\,,\ \psi(x_1)\,\tilde{\varphi}(x_2)\,\varphi(x_3)\,,\ 0\right)$$

and

$$\left(-\tilde{\psi}(x_1)\,\varphi(x_2)\left[\varphi(x_3+1) - \varphi(x_3)\right],\ 0\,,\ \psi(x_1)\,\varphi(x_2)\,\tilde{\varphi}(x_3)\right).$$

Let us explain why. Since $\alpha_1 = 1$, $\partial_1 : W^{(1)}_{j,\alpha} \to W_{j,\alpha}$ is an isomorphism. This means that we can choose a basis for the second component $u_2(x)$ together with a basis for the third component $u_3(x)$ and then compute $u_1(x)$. The basis which is chosen for the second component is $\psi(x_1 - k_1)\,\tilde{\varphi}(x_2 - k_2)\,\varphi(x_3 - k_3)$, $k_1 \in \mathbf{Z}$, $k_2 \in \mathbf{Z}$, $k_3 \in \mathbf{Z}$. This agrees with lemma 32 and with the conventions we made about V_j vs $\tilde{V}_j$. Similarly for the third one.

We now need to solve $\partial_1 u_1 + \partial_2 u_2 + \partial_3 u_3 = 0$. Let us begin with $u_2(x) = \psi(x_1)\,\tilde{\varphi}(x_2)\,\varphi(x_3)$ and $u_3(x) = 0$. Then $\frac{d}{dx_2}\,\tilde{\varphi}(x_2) = \varphi(x_2+1) - \varphi(x_2)$ while

$\tilde{\psi}(x_1)$ is precisely defined by $\frac{d}{dx_1}\tilde{\psi}(x_1) = \psi(x_1)$. That is explaining the first row.

The explanation of the second row is fully similar. If $\alpha = (1,1,0)$, the construction rules allow two prossibilities. We can either compute $u_1(x)$ when u_2 and u_3 are given or compute $u_2(x)$ in terms of u_1 and u_3.

We stick to the first option. It yields

$$(-\tilde{\psi}(x_1)\,\psi(x_2)\,\varphi(x_3)\,,\,\psi(x_1)\,\tilde{\psi}(x_2)\,\varphi(x_3)\,,\,0)$$

and

$$\Big(-\tilde{\psi}(x_1)\,\psi(x_2)\big(\varphi(x_3+1)-\varphi(x_3)\big)\,,\,0\,,\,\psi(x_1)\,\psi(x_2)\,\tilde{\varphi}(x_3)\Big)\,.$$

If $\alpha = (1,1,1)$ we also compute $u_1(x)$. It yields

$$(-\tilde{\psi}(x_1)\,\psi(x_2)\,\psi(x_3)\,,\,\psi(x_1)\,\tilde{\psi}(x_2)\,\psi(x_3)\,,\,0)$$

and

$$(-\tilde{\psi}(x_1)\,\psi(x_2)\,\psi(x_3)\,,\,0\,,\,\psi(x_1)\,\psi(x_2)\,\tilde{\psi}(x_3))\,.$$

The other cases are fully similar to these ones. We then have two mother wavelets inside each frequency channel. These frequency channels are indexed by $\alpha \in A$. Since $\sharp A = 7$, it yields 14 mother wavelets as announced.

28 Appendix 2. Wavelets and the div-curl lemma

P.L. Lions conjectured the following :

Let $E(x) = (E_1(x), E_2(x), \dots, E_n(x))$ and $B(x) = (B_1(x), B_2(x), \dots, B_n(x))$ be two vector fields satisfying the following three conditions

(28.1)	$E_j \in L^2(\mathbb{R}^n)$,	$B_j \in L^2(\mathbb{R}^n)$, $1 \le j \le n$
(28.2)	$\operatorname{div} E(x) = 0$	in the distributional sense
(28.3)	$\operatorname{curl} B(x) = 0$	in the distributional sense.

Then $(E \cdot B)(x) = E_1(x)B_1(x) + \cdots + E_n(x)B_n(x)$ belongs to the Stein & Weiss space $\mathcal{H}^1(\mathbb{R}^n)$.

My first reaction to P.L. Lions' conjecture was : this cannot be true ! It is so great that it should have been known if it were true.

But the following night I proved this conjecture and I acknowledged I was too pessimistic.

Here I want to provide the reader with a much better proof together with a systematic treatment of bilinear operators generalizing this specific example. This new approach is a joint work with Sylvia Dobyinsky [18].

We start with the heat semi-group $P_t = \exp(t\Delta)$, $\Delta = \frac{\partial^2}{\partial x_1^2} + \cdots + \frac{\partial^2}{\partial x_n^2}$. We then write $Q_t = -t\frac{\partial}{\partial t}P_t = -t\Delta P_t$. Then to any $f \in L^2(\mathbb{R}^n)$, we associate the function $g(x)$ defined as

$$(28.4) \qquad g(f)(x) = 2\left(\int_0^\infty |Q_t f(x)|^2 \frac{dt}{t}\right)^{1/2}$$

in full similarity with the celebrated Littlewood-Paley function.

A trivial calculation yields

$$(28.5) \qquad \int_{\mathbb{R}^n} |f(x)|^2\, dx = \int_{\mathbb{R}^n} g^2(x)\, dx\,.$$

We then consider the difference $h(x) = |f(x)|^2 - g^2(x)$. We obviously have $\int h(x)\, dx = 0$.

But much more is true. Indeed $h(x)$ belongs to the Stein & Weiss space $\mathcal{H}^1(\mathbb{R}^n)$ whenever $f \in L^2(\mathbb{R}^n)$. We need the bilinear version of this fact. The pseudo-product $u\sharp v$ between $u \in L^2(\mathbb{R}^n)$ and $v \in L^2(\mathbb{R}^n)$ is defined as

$$(28.6) \qquad u\sharp v(x) = 4\int_0^\infty Q_t(u)\, Q_t(v)\, \frac{dt}{t}\,.$$

We then have

Theorem 28.1. *There exists a constant $C(n)$ such that the difference between the product $u(x)\,v(x)$ and the pseudo-product $u\sharp v$ between two functions in $L^2(\mathbb{R}^n)$ belongs to $\mathcal{H}^1(\mathbb{R}^n)$ and satisfies*

$$(28.7) \qquad \|uv - u\sharp v\|_{\mathcal{H}^1} \le C(n)\, \|u\|_2\, \|v\|_2\,.$$

The proof of this fact is straightforward. A simple calculation yields

$$(28.8) \quad \begin{aligned} w(x) &= uv - u\sharp v \\ &= (2\pi)^{-2n} \iint \exp\big(ix\cdot(\xi+\eta)\big)\, \pi(\xi,\eta)\, \hat{u}(\xi)\, \hat{v}(\eta)\, d\xi\, d\eta \end{aligned}$$

where $\pi(\xi,\eta) = \left(\frac{|\xi|^2-|\eta|^2}{|\xi|^2+|\eta|^2}\right)^2$.

This bi-linear symbol π belongs to $C^\infty(\mathbb{R}^n \times \mathbb{R}^n \setminus \{0,0\})$. Moreover π is homogeneous of degree 0 and vanishes on $\xi + \eta = 0$. These three properties imply that the bilinear operator defined by (28.8) maps $L^2(\mathbb{R}^n) \times L^2(\mathbb{R}^n)$ into $\mathcal{H}^1(\mathbb{R}^n)$. A reference is [20] or [56].

We now return to the proof of P.L. Lions' conjecture. We need to prove the following

$$E_1(x)B_1(x) + \cdots + E_n(x)B_n(x) \in \mathcal{H}^1(\mathbb{R}^n) \tag{28.9}$$

whenever (28.1), (28.2) and (28.3) are satisfied.

We first replace each $E_j(x)B_j(x)$ by $E_j \sharp B_j$. The corresponding error term belongs to $\mathcal{H}^1(\mathbb{R}^n)$. We then treat the main terms by the following theorem.

Theorem 28.2. *If* (28.1), (28.2) *and* (28.3) *are satisfied, then the sum*

(28.10)
$$\sigma(x) = E_1\sharp B_1 + \cdots + E_n\sharp B_n \quad \textit{belongs to} \quad \dot{B}_1^{0,1}.$$

This homogeneous Besov space is defined by the following three equivalent properties

$$\sum_{-\infty}^{\infty} \|\Delta_j(f)\|_1 = \|f\|_{\dot{B}_1^{0,1}} < \infty \tag{28.11}$$

where $\Delta_j(f)$, $j \in \mathbf{Z}$, are the dyadic blocks of a Littlewood-Paley decomposition

$$f(x) = \sum\sum \alpha(j,k)\, 2^{nj}\, \psi(2^j x - k) \tag{28.12}$$

where $\sum\sum |\alpha(j,k)| < \infty$, $2^{nj/2}\,\psi(2^j x - k)$ being an orthonormal wavelet basis with enough smoothness and vanishing moments. Indeed $2^n - 1$ mother wavelets are needed and the sum over ψ is omitted.

The third definition of $\dot{B}_1^{0,1}$ is given by

$$\int_0^\infty \|Q_t(f)\|_1 \frac{dt}{t} < \infty. \tag{28.13}$$

For simplifying the notations, the norm in $\dot{B}_1^{0,1}$ will be denoted by $\|\cdot\|$. All other norms will carry indices. We have

$$\sigma(x) = 4\int_0^\infty Q_t(E)\cdot Q_t(B)\,\frac{dt}{t} \tag{28.14}$$

and by convexity of the ball of $\dot{B}_1^{0,1}$, it suffices to show the following

$$\int_0^\infty \|Q_t(E) \cdot Q_t(B)\| \frac{dt}{t} \leq C \|E\|_2 \|B\|_2 \tag{28.15}$$

when (28.1), (28.2) and (28.3) are satisfied.

For estimating $\|Q_t(E) \cdot Q_t(B)\|$, the following lemma will be used.

Lemma 37. *Let $f(x)$, $x \in \mathbb{R}^n$, be a function with the following two properties*

$$f = \operatorname{div} F \quad \textit{where} \quad F(x) = \big(F_1(x), \dots, F_n(x)\big) \tag{28.16}$$

and $\|F_j\|_1 \leq 1$, $1 \leq j \leq n$

$$\partial_j f \in L^1(\mathbb{R}^n) \quad \textit{and} \quad \|\partial_j f\|_1 \leq 1\,, \ 1 \leq j \leq n\,. \tag{28.17}$$

Then $\|f\| \leq C(n)$.

The proof is straightforward. If $\sum_{-\infty}^{\infty} \Delta_j(f)$ is the Littlewood-Paley expansion of f, we estimate $\|\Delta_j(f)\|_1$ by $2^j \|F\|_1$. This first estimate is obtained by the Bernstein's inequality applied to $\Delta_j(F)$. Bernstein's inequality reads

$$\|\partial_j u\|_1 \leq R \|u\|_1\,, \qquad 1 \leq j \leq n \tag{28.18}$$

when the Fourier transform of u is carried by $|\xi| \leq R$.

The second estimate is

$$\|\Delta_j(f)\|_1 \leq 2^{-j}\big(\|\partial_1 f\|_1 + \cdots + \|\partial_n f\|_1\big)\,. \tag{28.19}$$

It is obtained by writing

$$g = \Delta^{-1}\big(\partial_1(\partial_1 g) + \cdots + \partial_n(\partial_n g)\big) \tag{28.20}$$

and observing that

$$\|\Delta^{-1}(h)\|_1 \leq R^{-2} \|h\|_1 \tag{28.21}$$

if the Fourier transform of h vanishes on $|\xi| \leq R$.

Then Bernstein's inequality, (28.17) and (28.21) imply (28.19).

The conclusion of lemma can obviously be rewritten as

$$\|f\| \leq C(n) \left[\|\nabla f\|_1 + \|F\|_1\right]. \tag{28.22}$$

We now apply (28.22) to $f_t(x) = t^n f(tx)$ and observe that $\|f_t\| = \|f\|$. But $\|\nabla f_t\|_1 = t\,\|\nabla f\|_1$ while $\|F_t\|_1 = t^{-1}\,\|F\|_1$. By optimizing on $t > 0$, we obtain

$$\|f\| \le C(n)\big(\|\nabla f\|_1\,\|F\|_1\big)^{1/2}. \tag{28.23}$$

We now return to (28.15). The first remark is the following. Since $\operatorname{curl} B = 0$, we obtain

$$B(x) = \nabla U(x) \tag{28.24}$$

where U is scalar valued. Moreover $\|B_1\|_2 + \cdots + \|B_n\|_2$ and

$$\left(\int_0^\infty t^{-1}\|Q_t\,U\|_2^2\,\frac{dt}{t}\right)^{1/2} \tag{28.25}$$

are equivalent norms. Finally

$$Q_t(E)\cdot Q_t(B) = \operatorname{div} Q_t(E)\,Q_t(U)\,. \tag{28.26}$$

After writing (28.26), we can drop out the fundamental assumptions (28.2) and (28.3). Only (28.1) is retained.

Using the obvious inequality $2\sqrt{ab} \le t^{-1/2}a + t^{1/2}b$, $a, b > 0$, together with Cauchy-Schwarz, (28.23) implies

$$\|Q_t(E)\cdot Q_t(B)\| \le A(t) + B(t) + C(t) \tag{28.27}$$

where

$$\begin{aligned} A(t) &= C\,t^{-1/2}\,\|Q_t(U)\|_2\,\|Q_t(E)\|_2 \\ B(t) &= C\,t^{1/2}\sum_1^n\sum_1^n \|\partial_j Q_t E_k\|_2\,\|Q_t B_k\|_2 \\ \text{and}\qquad C(t) &= C\,t^{1/2}\sum_1^n\sum_1^n \|Q_t E_k\|_2\,\|\partial_j Q_t B_k\|_2\,. \end{aligned}$$

We want to prove that both $A(t)$, $B(t)$ and $C(t)$ can be written as $p(t)\,q(t)$ where

$$\left(\int_0^\infty p^2(t)\,\frac{dt}{t}\right)^{1/2} \le C\,\|E\|_2 \tag{28.28}$$

and

$$\left(\int_0^{\infty} q^2(t)\,\frac{dt}{t}\right)^{1/2} \le C\,\|B\|_2\,. \tag{28.29}$$

This will suffice to end our proof.

Concerning $A(t)$, we have $p(t) = t^{-1/2}\,\|Q_t(U)\|_2$, $q(t) = \|Q_t(E)\|_2$ and we simply use (28.25). The treatment of $B(t)$ is not deeper. We have $p(t) = t^{1/2}\|\partial_j Q_t E_k\|_2$ and $q(t) = \|Q_t B_k\|_2$. Indeed $t^{1/2}\,\partial_j\,Q_t\,f = \tilde{Q}_t f$ where

$$\int_0^{\infty} \|\tilde{Q}_t\|_2^2\,\frac{dt}{t} \le C\,\|f\|_2^2 \tag{28.30}$$

as Plancherel identity shows.

The treatment of $C(t)$ is similar and left to the reader.

Conclusion.

We return to our fundamental issue. What would we benefit from using a divergence-free wavelet basis ?

The proof would begin the same way. The vector field $E(x)$ is expanded into a divergence free wavelet basis

$$E(x) = \sum\sum\sum \alpha(j,k)\,2^{nj/2}\,\psi(2^j x - k) \tag{28.31}$$

where the first sum runs over the 14 divergence free mother wavelets, the second over all scales and the third over $k \in \mathbf{Z}^3$.

To expand the second vector field $B(x)$ we need to use a dual wavelet basis. This means that instead of $\tilde{V}_j$, $\tilde{W}_j$ we need to use $V_j^\sharp$, $W_j^\sharp$ where $W_0^\sharp$ is defined by the condition

$$\frac{d}{dt} : W_0 \to W_0^\sharp$$

is an isomorphism.

Similarly $\Phi(t-k)$, $k \in \mathbf{Z}$, is a Riesz basis of $V_0^\sharp$ where $\Phi(t-1) - \Phi(t) = \varphi'(t)$.

With these notations, $B(x)$ is expanded into the same type of basis that was used for $E(x)$. We then have

$$B(x) = \sum\sum\sum \beta(j,k)\,2^{nj/2}\,\psi^\sharp(2^j x - k) \tag{28.32}$$

where here there are only 7 mother wavelets. Indeed $B = \nabla U$ where $U(x)$ is a scalar valued function. Then (28.31) and (28.32) will be simplified into

$$E(x) = \sum_j \sum_\alpha E_{j,\alpha}(x) \tag{28.33}$$

$$B(x) = \sum_j \sum_\alpha B_{j,\alpha}(x) \tag{28.34}$$

and we write

$$E(x) \cdot B(x) = \sum_j \sum_\alpha E_{j,\alpha}(x) \cdot B_{j,\alpha}(x) + \Gamma(x)\,. \tag{28.35}$$

It can be proved that $\Gamma(x)$ belongs to $\mathcal{H}^1(\mathbb{R}^n)$. This is not related to $\operatorname{div} E(x) = 0$, $\operatorname{curl} B(x) = 0$. We now turn to the main term

$$\sigma(x) = \sum_j \sum_\alpha E_{j,\alpha}(x) \cdot B_{j,\alpha}(x)\,. \tag{28.36}$$

We know from Lemarié-Rieusset's clever construction that

$$\operatorname{div} E_{j,\alpha}(x) = 0 \quad , \quad \operatorname{curl} B_{j,\alpha}(x) = 0\,. \tag{28.37}$$

Then the proof we gave with a continuous formalism applies to this discrete formalism.

The conclusion is clear. Divergence-free wavelet bases do not pay.

29 Conclusion

We wanted to address a fundamental issue concerning the role played by wavelets or Littlewood-Paley analysis and paraproduct algorithms in solving Navier-Stokes equations.

There are indeed several issues. The first one concerns the mathematical theory of Navier-Stokes equations. The second issue which will be addressed is the role played by wavelets in the numerical simulation of Navier-Stokes equations.

The last problem addresses the role played by wavelets as a visualisation tool when one is working on experimental turbulence.

Let us first consider existence theorems. To the best of our understanding, the most powerful algorithm is the improvement on Kato's algorithm we developed with M.A. Muschietti.

This algorithm does not rely on wavelets or on Littlewood-Paley analysis. However if we want to apply this algorithm when the initial condition $u_0(x)$ is smooth, we are immediately entering the wavelet world. Indeed the largest amount of smoothness that can be imposed on $u_0(x)$ is defined by the minimal adapted Banach space E. This minimal Banach space is the homogeneous Besov space $\dot{B}_1^{2,1}(\mathbb{R}^3)$. This Banach space cannot be understood unless wavelet analysis is being used.

To be even more precise, let us assume that the initial condition $u_0(x)$ is a sum of a few normalized divergence-free wavelets. These wavelets are $2^j\psi(2^jx - k)$, $j \in \mathbf{Z}$, $k \in \mathbf{Z}^3$, $\psi \in A$. Here A is a finite set consisting of 14 mother wavelets $\psi = (\psi_1, \psi_2, \psi_3)$ where ψ_1, ψ_2 and ψ_3 belong to the Schwartz class.

The coefficients of this expansion of $u_0(x)$ are denoted by $\alpha(j,k)$, $j \in \mathbf{Z}$, $k \in \mathbf{Z}^3$, and are assumed to belong to $\ell^1(\mathbf{Z}^4)$. Then this property of u_0 is preserved by the Navier-Stokes equations : uniformly in $t \geq 0$, $x \to u(x,t)$ is enjoying the same regularity whenever $\sum_j \sum_k |\alpha(j,k)| \leq \eta$.

These remarks imply that P. Federbush was right. Indeed "Navier and Stokes meet the wavelet" as it was announced by P. Federbush. But this meeting did not take place in the room which was arranged by P. Federbush. The Morrey-Campanato space $M^2(\mathbb{R}^3)$ is the largest space which is adapated to Navier-Stokes equations and P. Federbush failed in using a wavelet based Galerkin algorithm to construct mild solutions when $u_0(x) \in M^2(\mathbb{R}^3)$. Such a construction was given in these notes using Kato's approach to the problem.

We should remark that functions in $M^2(\mathbb{R}^3)$ admit a simple characterization by size properties of wavelet coefficients. However this characterization is not the best approach to $M^2(\mathbb{R}^3)$ which is more naturally defined by plain size estimates. This could be the reason why Kato's algorithm is so efficient in this situation while P. Federbush failed. It is possible that Federbush's approach would work in the context of the minimal space $\dot{B}_1^{2,1}(\mathbb{R}^3)$ for which a wavelet analysis is so natural.

Let us then turn to uniqueness of mild solutions. Here also we have to face bad news. It is clear that the original proof by P.G. Lemarié-Rieusset and his collaborators is based on Besov spaces and paraproduct algorithms. But this proof nowadays appears as awkward as compared to the proof based on the "real variable methods" developed by A. Calderón and A. Zygmund.

Does it mean that real variable methods are always winning against spectral methods when applied to Navier-Stokes equations ?

This is certainly not the case. Indeed mathematicians are incorporating wavelet analysis inside a broader context which is named "micro-local analysis".

A.P. Calderón announced in 1965 that a better understanding of algebras of pseudo-differential coefficients with minimal smoothness assumptions should have a tremendous impact on non-linear PDE's.

Then Calderón proposed a collection of entirely new operators. It was a challenging task to prove the boundedness of these operators. Today the best proofs are based on a wavelet analysis.

This means that wavelet analysis is obviously playing an important role in Calderón's program.

J.M. Bony reshaped Calderón's program and launched the celebrated para-differential calculus. The goal is the same as in Calderón's approach. It consists in including in the calculus the pointwise multiplications by non-smooth functions.

As indicated in [56], there exists a version of the paraproduct operator $\pi(a, f)$ which is diagonal in a wavelet basis.

This does not mean that wavelets are playing a key role in Bony's program. However Bony's program is not antagonistic to wavelet analysis.

Bony's para-differential operators have important applications to many problems concerning Navier-Stokes equations. The best references are J.Y. Chemin [16], T. Kato and G. Ponce and finally Michael Taylor [61], [62].

The main ingredients in these approaches are non trivial commutator estimates. These estimates are similar to the ones Calderón, Coifman and myself were able to prove. Let us stress that the commutators which are involved lie beyond pseudo-differential operators.

Returning to the main issue, should one bet on wavelets for a deeper insight into Navier-Stokes equations ? Let us confess that we do not know. In contrast we can already state that paradifferential operators or microlocal analysis are already playing a key role. The first person who tried to convince me that microlocal analysis and non-linear PDE's had strong ties was Luc Tartar. But at that time my knowledge and my interest in non-linear PDE's was too limited to understand his deep views.

We now turn to a distinct issue which is the role that wavelets might play in numerical simulations of Navier-Stokes equations.

Many scientists are hoping that wavelets will be a key ingredient in the next century codes. A description of this research area is extracted from a research program by Nicholas Kevlahan, LMD, ENS, 24 rue Lhomond, 75231 Paris Cedex 05.

> "The wavelet transform was first introduced as an analysis technique, but numerical methods have been developed recently which use wavelet bases to actually solve partial differential equations (Frölich & Schneider 1996 ; Charton & Perrier 1996). These methods are particularly well-suited to equations, such as the Navier-Stokes equations at high Reynolds number, whose solutions contain isolated multiscale structures or quasi-singularities. In collaboration with Kai Schneider (ICT, Universität Karlsruhe) I compared simulations using these wavelet techniques with standard spectral simulations and nonlinearly filtered spectral simulations (Schneider, Kevlahan & Farge 1997). The results showed that the wavelet methods are very accurate, and require fewer active modes than spectral methods. Furthermore, the number of active wavelet modes is approximately constant in time, even during intense nonlinear interactions, whereas the number of active spectral modes peaks when the interactions are most intense..."

In his Ph. D. dissertation, Mats Holmström also advocates wavelets for solving PDEs.

> "In fluid dynamics we have shocks, boundary layers and turbulence. For these examples the solution can be smooth in most of the solution domain, with small areas where the solution changes quickly. When solving such problems numerically we would like to adjust the discretization to the solution. In terms of finite difference methods, we want to have many points in areas where the solution has strong variation, and few points where the solution is smooth. If we use a Galerkin method this corresponds to the representation of the solution having fewer basis functions in the smooth areas. ... The most common way of compressing such a representation is thresholding. We delete all wavelet coefficients of magnitude less than some threshold ε ... Note that by thresholding a wavelet representation we have a way to automatically find a sparse representation, and we can

> also use this representation to compute function values at any point...
>
> The method can be viewed as an adaptive mesh method, where the mesh is automatically refined around sharp variations of the solution. An advantage of the method is that we never have to care about where and how to refine the mesh. All this is handled by thresholding the wavelet coefficients. The method is well suited for large problems with a solution that is well compressed in a wavelet basis... The constants in the estimates are large. They can be reduced by using other wavelets than Daubechies wavelets but the problem size has to be large before the wavelet method outperforms classical methods..."

We strongly believe in wavelet based algorithms for solving non linear PDEs. When more powerful computers are available, wavelet based algorithms might win. Indeed the issue will then be a sharper analysis of the shocks or singularities that might develop in the non-linear evolution.

Let me however qualify this remark. In my opinion, wavelet analysis belongs to the same group of tools as refinements of meshes or multipanel processing do. We also would like to include the celebrated Oslo algorithm ("knot removal") in this group. All these algorithms are addressing the same issue : one should adapt the grid to the solution which is being computed. In the case of an evolution equation, the grid should also evolve in time.

The equivalence between wavelet analysis, refinements of meshes and multipanel processing is grounded on the pioneering work on non-linear approximation which was achieved by De Vore and inspired by Peller and Peetre.

Returning to Navier-Stokes equations, my belief concerns algorithms that mimic wavelet methods. These algorithms automatically introduce more segmentation or grid points when a singularity is developing.

A last issue concerns the role played by wavelets as a visualisation tool.

Let us return to N. Kevlahan. He is advocating a wavelet analysis as a post-processing :

> "The wavelet transform is a new harmonic analysis technique developed in France during the 1980s (Grossmann & Morlet). Marie Farge showed that this technique is appropriate for the analysis of turbulence because it permits a localization that is

both spatial and spectral ([27], [30]). This means that one can link the physical characteristics of turbulence (e.g. the presence of vortices) to its pectral or statistical properties (e.g. energy spectrum slope) [28]."

This means that here wavelets are used as a visualisation tool.

Then N. Kevlahan confirms M. Farge's hopes :

"A wavelet analysis permitted me to cleanly separate the vorticity filaments from the coherent vortices..."

This was achieved on a numerical simulation of two-dimensional turbulence. The simulation was obtained with a conventional algorithm, but wavelets were used as a discrimination tool in a pattern analysis. The success of the work of Farge and Kevlahan responds to a concern expressed by R. Azencott that detecting and isolating coherent vortices is a highly non trivial task and that wavelet people might be using the wrong tools.

While M. Farge and his group are using a wavelet analysis for detecting and isolating vertices on numerical simulations of 2D turbulence, A. Arneodo and his team used a wavelet analysis on the turbulence signal which was measured by Y. Gagne inside the Modane wind tunnel.

A. Arneodo and his team made two important discoveries. He could prove the multifractal structure of the velocity field as a function of the time variable [3] and he was able to detect huge transients which were interpreted by A. Arneodo as vorticity filaments crossing the hot wire. This hot wire is used to measure the velocity inside the wind tunnel.

Arneodo's hypothesis was eventually confirmed by Y. Couder and his group [26].

It is time to conclude.

It is now obvious that (1) wavelet analysis is the best available tool for scrutinizing the intricate behavior of experimental turbulence and (2) there are some hopes that wavelet type algorithms will be increasingly used in numerical simulations. I am including adaptive refinement of meshes and multipanel processing inside this group of algorithms.

Bibliography

[1] P. Abry. *Ondelettes et turbulences.* Diderot ed. (1997).

[2] S. Alinhac and P. Gérard. *Opérateurs pseudo-différentiels et théorème de Nash-Moser.* Savoirs actuels. CNRS (1991).

[3] A. Arneodo, F. Argoul, E. Bacry, J. Elezgaray & J.F. Muzy. *Ondelettes, multifractales et turbulences.* Diderot ed. (1995).

[4] G. Battle and P. Federbush. *Divergence-free vector wavelets.* Michigan Math. Journal. **40** (1993) 181–195.

[5] O. Barraza. *Self-similar solutions in weak L^p spaces of the Navier-Stokes equations.* Rev. Mat. Iberoamericana. Vol. 12 no. 2 (1996) 411–439.

[6] S. Bertoluzza. *An adaptive collocation method based on interpolating wavelets.* In : W. Dahmen, A.J. Kurdila and P. Oswald, eds. Multiscale Wavelet Methods for PDE's. Academic Press, New York (1997).

[7] G. Beylkin and J.M. Keiser. *An adaptive pseudo-wavelet approach for solving nonlinear partial differential equations.* In : W. Dahmen, A.J. Kurdila and P. Oswald eds. Multiscale Wavelet Methods for PDE's. Academic Press, New York (1997).

[8] J.M. Bony. *Calcul symbolique et propagation des singularités pour les équations aux dérivées partielles non linéaires.* Ann. Sci. E.N.S. (1981), 209–246.

[9] J.H. Bramble, J.E. Pasciak and J. Xu. *Parallel multilevel preconditioners.* Math. Comp. **55** (1990) 1–22.

[10] L. Caffarelli, R. Kohn and L. Nirenberg. *Partial regularity of suitable weak solutions of the Navier-Stokes equations.* Comm. Pure Appl. Math. **35** no. 6 (1982) 771–831.

[11] A.P. Calderón. *Algebras of singular integral operators.* AMS Proc. Symp. Pure Math X (1967) 18–55.

[12] M. Cannone. *Ondelettes, paraproduits et Navier-Stokes.* Diderot ed. (1995).

[13] M. Cannone, Y. Meyer and F. Planchon. *Solutions auto-similaires des équations de Navier-Stokes.* Séminaire EDP. Ecole Polytechnique. F–91128 Palaiseau Cedex (18 janvier 1994).

[14] M. Cannone and F. Planchon. *Self-similar solutions for Navier-Stokes equations in* $\mathbb{R}^3$. Comm. Partial Differential Equations **21** no. 1–2 (1995) 179–193.

[15] P. Charton and V. Perrier. *A pseudo-wavelet scheme for the two-dimensional Navier-Stokes equation.* Comp. Appl. Math. **15** (2) (1996) 139–160.

[16] J. Chemin. *Calcul paradifférentiel précisé et applications ê des équations aux dérivées partielles non linéaires.* Duke Math. J. **56** (1988) 431–469.

[17] A. Cohen, *Wavelets methods in numerical analysis.* Handbook of Numerical Analysis. P.G. Ciarlet and J.L. Lions. Techniques of scientific computing (1998).

[18] R.R. Coifman, S. Dobyinsky and Y. Meyer. *Opérateurs bilinéaires et renormalisation.* In : Essays on Fourier analysis in honor of Elias M. Stein ed. by Ch. Fefferman, R. Fefferman and S. Wainger (1995).

[19] R. Coifman, P.L. Lions, Y. Meyer and S. Semmes. *Compensated compactness and Hardy spaces.* J. Math. Pures et Appl. (**9**) 72 no. 3 (1993) 247–286.

[20] R. Coifman and Y. Meyer. *Au delê des opérateurs pseudo-différentiels.* Astérisque 57. Soc. Math. de France (1978).

[21] S. Dahlke and I. Weinreich. *Wavelet-Galerkin methods : an adapted biorthogonal wavelet basis.* Constr. Approx. **9** (1993) 237–262.

[22] W. Dahmen. *Wavelets and multiscale methods for operator equations.* Acta Numerica. Cambridge University Press (1997) 55–228.

[23] W. Dahmen and A. Kunoth. *Multilevel preconditioning.* Numer. Math. **63** (1992) 315–344.

[24] W. Dahmen, A. Kunoth and K. Urban. *A wavelet Galerkin method for the Stokes problem.* Computing **56** (1996) 259–302.

[25] S. Dobyinski. *La version ondelettes du théorème du Jacobien.* Rev. Mat. Iberoamericana **11** (2) (1995) 309–333.

[26] S. Douady, Y. Couder and M. Brachet. *Direct observation of the intermittency of intense vorticity filaments in turbulence.* Phys. Rev. Lett. **67** (1991) 983–986.

[27] M. Farge. *Wavelet transforms and their applications to turbulence.* Ann. Rev. Fluid Mech. **24** (1992) 395–457.

[28] M. Farge, N. Kevlahan, V. Perrier and E. Goirand. *Wavelets and turbulence.* Proceedings of the IEEE **84** (4), April 1996, 639–669.

[29] M. Farge, N. Kevlahan, V. Perrier and K. Schneider. *Turbulence analysis, modelling and computing using wavelets.* In : Wavelets and Physics. (ed. H. van der Berg). Cambridge University Press (1997).

[30] M. Farge and G. Rabreau. *Transformée en ondelettes pour détecter et analyser les structures cohérentes dans les écoulements turbulents bidimensionnels.* C.R. Acad. Sci. Paris Ser. II **307** (1988) 433–462.

[31] P. Federbush. *Navier and Stokes meet the wavelet.* Commun. Math. Phys. **155** (1993) 219–248.

[32] C. Fefferman and E. Stein. *H^p functions of several variables.* Acta Math. **129** (1972) 137–193.

[33] P. Frick and V. Zimin. *Hierarchical models of turbulence. Wavelets, Fractals and Fourier transforms.* Farge et al. eds. New York, Clarendon (1993) 265–283.

[34] U. Frisch. *Turbulence : the legacy of A.N. Kolmogorov.* Cambridge Univ. Press (1995).

[35] G. Furioli, P.G. Lemarié-Rieusset and E. Terraneo. *Sur l'unicité dans $L^3(\mathbb{R}^3)$ des solutions "mild" des équations de Navier-Stokes.* Note aux CRAS. A para"tre.

[36] W. Hackbush. *Multigrid methods and applications.* Springer Verlag, New York (1985).

[37] A. Harten. *Adaptive multiresolution schemes for shock computations.* Journal of Comp. Physics **115** (2) (1994) 319–338.

[38] A. Harten. *Multiresolution algorithms for the numerical solution of hyperbolic conservation laws.* Comm. Pure Appl. Math. **48** (1995) 1305–1342.

[39] M. Holmstrm. *Wavelet based methods for time dependent PDEs.* Uppsala Univ. (1997).

[40] S. Jaffard. *Wavelet methods for fast resolution of elliptic problems.* SIAM J. Numer. Anal. **29** (1992) 965–986.

[41] J.P. Kahane. *Séries de Fourier absolument convergentes.* Springer Verlag (1970).

[42] J.P. Kahane and P.G. Lemarié-Rieusset. *Fourier series and wavelets.* Gordon and Breach publishing group (1996).

[43] T. Kato. *Strong L^p-solutions of the Navier-Stokes equations in $\mathbb{R}^m$ with applications to weak solutions.* Math. Z. **187** no. 4 (1984) 471–480.

[44] T. Kato and G. Ponce. *Well-posedness of the Euler and Navier-Stokes equations in the Lebesgue spaces L^p_s.* Rev. Mat. Iberoamericana **2** no. 1–2 (1984) 73–88.

[45] T. Kato and G. Ponce. *Commutator estimates and the Euler and Navier-Stokes equations.* CPAM **41** (1988) 891–907.

[46] N. Kevlahan. *Research program : From coherent structures to a statistical theory of turbulence.* LMD, ENS, 24 rue Lhomond, 75231 Paris Cedex 05.

[47] A. Kunoth. *Multilevel Preconditioning.* Verlag Shaker Aachen (1994).

[48] A. Kunoth. *Multiscale Preconditioning.* Advances in Comp. Math. **4** (1995) 145–170.

[49] S. Lazaar, J. Liandrat et P. Tchamitchian. *Algorithme ê base d'ondelettes pour la résolution numérique d'équations aux dérivées partielles ê coefficients variables.* CRAS Paris. Série I. **319** (1994) 1101–1107.

[50] P.G. Lemarié. *Analyses multirésolutions non-orthogonales, commutation entre projecteurs et dérivation et ondelettes vecteurs ê divergence nulle.* Rev. Mat. Iberoamericana **8** (1992) 221–236.

[51] J. Leray. *Essai sur le mouvement d'un liquide visqueux emplissant l'espace.* Acta Math. **63** (1934) 193–248.

[52] T. Lyche. *Knot removal for spline curves and surfaces.* In : E.W. Cheney, C.K. Chui and L. Schumaker eds. Approximation Theory VII. Academic Press, Boston (1993) 207–227.

[53] Y. Meyer. *Remarques sur un théorème de J.M. Bony.* Rend. del Circulo mat. di Palermo (suppl. II, 1) (1981) 1–20.

[54] Y. Meyer. *Wavelets and fast numerical algorithms.* Handbook of Numerical Analysis, Vol. V, ed. by P.G. Ciarlet and J.L. Lions (1997) Elsevier Science B.V.

[55] Y. Meyer. *Wavelets and operators.* Cambridge studies in advanced mathematics **37** (1992).

[56] Y. Meyer and R. Coifman. *Calderón-Zygmund and multilinear operators.* Cambridge studies in advanced mathematics **48** (1997).

[57] F. Oru. Thèse (1998, Ecole Polytechnique, Centre de Mathématique).

[58] E.M. Stein. *Singular integrals and differentiability properties of functions.* Princeton Univ. Press (1970).

[59] L. Tartar. *Nonlinear partial differential equations using compactness method.* M.C.R. report 1584, University of Wisconsin (1976).

[60] L. Tartar. *Compensated compactness and applications to partial differential equations.* Nonlinear analysis and mechanics. Heriot-Watt Symposium. Vol. IV. Res. Notes in Math. Vol. 39, Pitman (1979) 136–212.

[61] M.E. Taylor. *Pseudodifferential operators and nonlinear PDE.* Birkhuser (1991).

[62] M.E. Taylor. *Analysis on Morrey spaces and applications to Navier-Stokes and other evolution equations.* Commun. in Partial Diff. Equations **17** (1992) 1407–1456.

[63] M.E. Taylor. *Partial differential equations III. Nonlinear equations.* Applied Mathematical Sciences **117** Springer-Verlag (1996).

[64] Ph. Tchamitchian. *Inversion explicite de certains opérateurs elliptiques.* SIAM J. Math. Anal. **27** (1996) 1680–1703.

[65] R. Temam. *Navier-Stokes equations. Theory and numerical analysis.* North-Holland Publishing Co. Amsterdam-New York (1984).

[66] O.V. Vasilyev and S. Paolucci. *A dynamically adaptive multilevel wavelet collocation method for solving partial differential equations in a finite domain.* Journal of Comp. Physics **120** (1) (1995) 33–47.

[67] H. Yserentant. *Hierarchical bases.* Proc. ICIAM 91 (1991) 256–276.

Made in United States
Orlando, FL
04 November 2026

90206100R00122